Wolfgang Hackbusch
Kristian Witsch (Eds.)

Numerical Techniques in Continuum Mechanics

Notes on Numerical Fluid Mechanics
Volume 16

Wolfgang Hackbusch
Kristian Witsch (Eds.)

Numerical Techniques in Continuum Mechanics

Proceedings of the Second GAMM-Seminar,
Kiel, January 17 to 19, 1986

Friedr. Vieweg & Sohn Braunschweig / Wiesbaden

CIP-Kurztitelaufnahme der Deutschen Bibliothek

Numerical techniques in continuum mechanics:
Kiel, January 17 to 19, 1986/Wolfgang Hackbusch;
Kristian Witsch (ed.). – Braunschweig; Wiesbaden:
Vieweg, 1987.
(Proceedings of the ... GAMM seminar; 2)
(Notes on numerical fluid mechanics; Vol. 16)
ISBN 3-528-08091-4

NE: Hackbusch, Wolfgang [Hrsg.]; Gesellschaft für
Angewandte Mathematik und Mechanik: Proceedings
of the ...; 2. GT

Manuscripts should have well over 100 pages. As they will be reproduced photomechanically they should be typed with utmost care on special stationary which will be supplied on request. In print, the size will be reduced linearly to approximately 75 %. Figures and diagrams should be lettered accordingly so as to produce letters not smaller than 2 mm in print. The same is valid for handwritten formulae. Manuscripts (in English) or proposals should be sent to the general editor Prof. Dr. E. H. Hirschel, Herzog-Heinrich-Weg 6, D-8011 Zorneding.

The addresses of the editors of the series are given on the inner back cover.

Produced by W. Langelüddecke, Braunschweig
Printed in Germany

ISSN 0179-9614

ISBN 3-528-08091-4

Foreword

The GAMM Committee for Efficient Numerical Methods for Partial Differential Equations (GAMM-Fachausschuß "Effiziente numerische Verfahren für partielle Differenzialgleichungen") organizes conferences and seminars on subjects concerning the algorithmic treatment of partial differential equation problems.

The first seminar "Efficient Solution of Elliptic Systems" was followed by a second one held at the University of Kiel from January 17th to January 19th, 1986. The title was

"Efficient Numerical Methods in Continuum Mechanics".

The equations arising in continuum mechanics have many connections to those of fluid mechanics, but are usually more complex. Therefore, much attention has to be paid to the efficient discretization, postprocessing and extrapolation.

The seminar was attended by 66 scientists from 10 countries. Most of the 21 lectures presented at the seminar treated the discretization of equations in continuum mechanics by finite elements, methods for improving the accuracy of these approximations and the use of boundary elements. Other contributions presented efficient methods for investigating bifurcations which play an essential role in practical applications. These proceedings contain 11 contributions in alphabetical order.

The editors and organizers of the seminar would like to thank the land Schleswig-Holstein and the DFG (Deutsche Forschungsgemeinschaft) for their support.

Kiel, November 1986

W. Hackbusch
K. Witsch

Contents

OPTIMAL ERROR ESTIMATES AND ADAPTIVE TIME AND SPACE STEP CONTROL FOR LINEAR PARABOLIC PROBLEMS

by

Kenneth Eriksson, Claes Johnson and Johan Lennblad

Chalmers University of Technology and the University
of Göteborg, Department of Mathematics
S-412 96 GÖTEBORG, Sweden

0. Introduction

In this note we present recent developments in the program for constructing adaptive algorithms for numerical methods for parabolic type problems or stiff initial value prolems that was initiated in Johnson [6] and was continued in Eriksson, Johnson [3] and Johnson, Nie, Thomée [7].

Solutions of parabolic problems typically are nonsmooth in initial transients but become smoother as time increases. To minimize the work required to compute an approximate solution of a parabolic problem to a certain accuracy one therefore would like to use a numerical method which automatically adapts the mesh size (in time and space) according to the smoothness of the exact solution and automatically chooses a fine mesh in a transient and increases the mesh size as the exact solution becomes smoother. Our objective is to construct such adaptive algorithms that in particular satisfy the following criteria:

The error in the approximate solution is controlled globally in time to a given tolerance. (0.1)

The algorithm is efficient in the sense that the mesh size is not chosen unnecessarily small. (0.2)

The extra work required for the mesh control is small. (0.3)

The algorithm can be theoretically justified. (0.4)

No, or only very rough, a priori information of the exact solution is required. (0.5)

To be able to satisfy (0.5), the necessary information concerning the smoothness of the exact solution (of course) must be obtained from the computed approximate solution as the computation proceeds.

In [7] this program was carried out in detail in the particular case of a backward Euler semi-discretization in time of a linear parabolic problem with error control in the L_2-norm in space. We shall in this paper present extensions of these results to a fully discrete linear parabolic problem with now discretization also in space and with a higher order (third order) accurate method for the time discretization. The presented algorithm is easy to implement, satisfies (0.1) - (0.5) and seems according to our numerical tests to perform very satisfactory in practice. We believe that this type of algorithm may be very useful in applications. Extensions to non-linear parabolic problems will be presented in subsequent work.

Our discretization method is obtained by using a standard finite element method in space and the discontinuous Galerkin method in time. We consider in this paper the case of a piecewise linear approximation in time resulting in a third order accurate implicit Runge-Kutta type time-stepping scheme. Note that the backward Euler method considered in [3] corresponds to the discontinuous Galerkin method with piecewise constants. Our adaptive method is based on an *a posteriori* error estimate involving only the computed approximate solution. This estimate is obtained through an optimal *a priori* error estimate involving the unknown exact solution together with a result showing that under reasonable assumptions the quantities depending on the exact solution may be estimated using the computed approximate solution. In this note the adaptivity in space is restricted so that we only allow space meshes that become coarser as time increases. As indicated this covers the standard situation where the exact solution becomes smoother with increasing time. However, with given heat production terms or boundary conditions varying rapidly in time, reverse situations may occur. Such cases may be handled by the present technique through restart but would otherwise require a (non-obvious) extension of the argument.

We assume in this paper that the space discretization is quasi-uniform on each time level and thus the local element size in space only depends on time. It is conceivable to allow also a dependence on the space variable and thus work with space meshes refined locally in space. The control of such local refinements will however require local error estimates the proof of which will involve additional technical complications. A first step towards adaptive local refinements for elliptic problems based on local error estimates was taken in [5]. We hope to be able to extend this type of results to parabolic problems in future work.

An outline of this note is as follows. In Section 1 we introduce the fully discrete numerical method and state the optimal a priori error estimate. In Section 2 we formulate the associated adaptive algorithm, and finally, in Section 3 we present the results of some numerical experiments. For a proof of the a priori error estimate, we refer to [4]. The proof of the a posteriori error estimate, which is analogous to a corresponding proof in [7], will be given in a future publication.

For a more detailed comparison (as concerns the time discretization) of adaptive methods of the type considered in this note with earlier methods presented in the literature for numerical methods for stiff systems of ordinary differential equations, we refer to the discussion in [6]. Let us here just remark that with the earlier approach to adaptivity for stiff problems it seems as if one faces serious difficulties with respect to all the conditions (0.1)-(0.4).

1. Discretization and a priori estimates

As a model problem we shall consider the following parabolic problem: Find $u:(0,\infty) \to H^2(\Omega) \cap H_0^1(\Omega)$ such that

$$\begin{aligned} u_t - \Delta u &= f \quad \text{in} \quad \Omega, \ t > 0, \\ u(0) &= u_0 \quad \text{in} \quad \Omega, \end{aligned} \tag{1.1}$$

where Ω is a bounded domain in R^d with smooth boundary Γ, u_0 and f are given data and $u_t = \frac{\partial u}{\partial t}$. Here and below

$H^s(\Omega)$ denotes for $s \geq 0$ the usual Sobolev space (of functions with derivatives of order s square integrable over Ω) with norm $\|\cdot\|_s$ and corresponding semi-norm $|\cdot|_s$ and $H_0^1(\Omega) = \{v \in H^1(\Omega): v = 0$ on $\Gamma\}$. As is well-known, (1.1) may be given the following variational formulation: Find $u:(0,\infty) \to H_0^1(\Omega)$ such that

$$(u_t,v) + (\nabla u, \nabla v) = (f,v) \quad \forall v \in H_0^1(\Omega),\ t > 0 \quad (1.2)$$

and $u(0) = u_0$, where $(.,.)$ denotes the $L_2(\Omega)$-inner product.

To discretize (1.2) let $0 = t_0 < t_1 <\ldots< t_n <\ldots,$ be a subdivision of $(0,\infty)$ into time intervals $I_n = (t_{n-1}, t_n]$ of length $k_n = t_n - t_{n-1}$ and let $S_n \subset H_0^1(\Omega)$, $n = 1,2,\ldots,$ be finite dimensional spaces satisfying for some $r \geq 2$ and constant $\bar{C}$,

$$\inf_{\psi \in S_n} \|\varphi - \psi\|_j \leq \bar{C} h_n^{r-j} |\varphi|_r, \quad j = 0,1,\ \forall \varphi \in H^r(\Omega). \quad (1.3)$$

Here the S_n are typically finite element spaces based on continous piecewise polynomial functions of degree at most $r - 1$ on quasi-uniform triangulations of Ω with mesh size h_n. For a given non-negative integer q we introduce for $n = 1,2,\ldots,$ the finite dimensional space V_n consisting of functions on I_n with values in S_n that vary as polynomials of degree at most q in time:

$$V_n = \{v: I_n \to S_n: v(t) = \sum_{j=0}^{q} t^j a_j,\ a_j \in S_n\}.$$

We shall seek an approximate solution U in the space V defined by

$$V = \{v:(0,\infty) \to H_0^1(\Omega): v|_{I_n} \in V_n,\ n = 1,2,\ldots,\}.$$

To account for the fact that the functions in V may be discontinuous in time at the discrete time levels t_n, we introduce the notation

$$v_n^{\pm} = \lim_{s \to 0^{\pm}} v(t_n + s).$$

We shall consider the following numerical method for (1.2): Find $U \in V$ such that for $n = 1,2,\ldots,$ $U \equiv U|_{I_n}$ satisfies

$$\int_{I_n} \{(U_t,v) + (\nabla U,\nabla v)\}dt + (U_{n-1}^+ - U_{n-1}^-, v_{n-1}^+) =$$

$$\int_{I_n} (f,v)dt \quad \forall\, v \in V_n, \qquad (1.4)$$

where $U_0^- = u_0$. Note that with $q = 0$, (1.4) reduces to the following method: For $n = 1,2,\ldots,$ find $U_n \equiv U_n^-$ S_n such that

$$(U_n - U_{n-1},v)+k_n(\nabla U_n,\nabla v) = (\int_{I_n} f(t)dt,v) \;\forall v \in S_n, \quad (1.5)$$

which is a variant of the well-known backward Euler method where the average over I_n of the right hand side f is used instead of the usual value $f(t_n)$. Further, for $q = 1$ we get the following method

$$(\Psi_n,v) + k_n(\nabla\phi_n, v) + \frac{k_n}{2}(\nabla\Psi_n,\nabla v) + (\phi_n,v)$$

$$= (\phi_{n-1} + \Psi_{n-1},v) + (\int_{I_n} fdt,v) \qquad (1.6)$$

$$\frac{1}{2}(\Psi_n,w) + \frac{k_n}{2}(\nabla\phi_n,\nabla w) + \frac{k_n}{3}(\nabla\Psi_n,\nabla w)$$

$$= (\frac{1}{k_n}\int_{I_n} (t-t_{n-1})fdt,w)$$

$$\forall v,w \in S_n$$

where

$$U|_{I_n} = \phi_n + \frac{t-t_{n-1}}{k_n}\Psi_n, \quad \phi_n, \Psi_n \in S_n.$$

If $f = 0$, then this method for the time discretization corresponds to the subdiagonal Padé method of order 3, see [2].

In this paper we shall consider (1.4) with $q = 1$, that is, the method (1.6). The a priori estimate on which the adaptive algorithm for (1.6) is based reads as follows. We use the notation

$$\|v\|_{s,I_n} = \max_{t\in I_n} \|v(t)\|_s, \quad \|v\|_{I_n} \equiv \|v\|_{0,I_n}, \quad \|v\| \equiv \|v\|_0,$$

and

$$L_N = (\log\frac{t_N}{k_N} + 1)^{1/2}.$$

By C and C_i, $i = 0,\dots,3$, we denote positive constants only depending on the parameter γ and the constant $\bar{C}$ in (1.3).

Theorem 1. Let u be the solution of (1.1) and U that of (1.4) with $q = 1$. Suppose that $S_n \subseteq S_{n-1}$ for $n = 2,3,\dots,$ and that for some constant $\gamma > 1$ the time steps satisfy $\gamma k_n \le t_N - t_{n-1}$ for $1 \le n < N$, $N = 1,2,\dots$. Then we have for $N = 1,2,\dots$,

$$\|u-U\|_{I_N} \le L_N \max(C_0 h_n^r |u|_{r,I_n} + \min(C_1 k_n \|u_t\|_{I_n}, C_2 k_n^2 \|u_{tt}\|_{I_n})). \tag{1.7}$$

Theorem 2. Under the assumptions of Theorem 1 one has for $N = 1,2,\dots,$

$$\|u(t_N) - U_N^-\| \le L_N \max_{n \le N}(C_0 h_n^r \|u\|_{r,I_n} \tag{1.8}$$
$$+ \min(C_1 k_n \|\Delta u\|_{I_n}, C_2 k_n^2 \|\Delta u_t\|_{I_n}, C_3 k_n^3 \|\Delta u_{tt}\|_{I_n})).$$

It follows that the method (1.4) with $q = 1$ is second order accurate globally in time and third order accurate at the discrete time levels t_N. We also note that (1.7) is, disregarding the logarithmic factor, optimal in the sense that for some positive constant c

$$\inf_{v \in V} \|u-v\|_{(0,t_N)} \ge c \max_{n \le N} (h_n^r |u|_{r,I_n} + k_n^2 \|u_{tt}\|_{I_n}).$$

Clearly also (1.8) is optimal in the sense that we cannot increase the exponents of the factors h_n^r and k_n^3, neither can we use weaker norms on u while keeping the exponents of h_n and k_n. The adaptive algorithm to be introduced will be based on (1.8). The optimality of (1.8) will guarantee that condition (0.2) will be satisfied.

Remark 1.1. In general the minimum on the right hand side of (1.8) will be given by the third order tem $C_3 k_n^3 \|\Delta u_{tt}\|_{I_n}$. However, for the very few first steps instead the first order term $C_1 k_n \|\Delta u\|_{I_n}$ may give the minimum. Note that in the case $f \equiv 0$ this term may be replaced by

$$C_1 \int_{I_n} \|u_t(s)\| ds,$$

which is useful on the first interval where u_t may be unbounded. □

Remark 1.2. Note that the only constant in (1.7) and (1.8) depending on t_N is the logarithmic constant L_N. This means that it is possible to integrate over long time-intervals essentially without accumulation of errors. This reflects the parabolic nature of our problem. □

2. The adaptive algorithm

Suppose $\delta > 0$ is a given tolerance and that we want the error $e = u-U$ in the approximate solution given by (1.4) with $q = 1$ to satisfy

$$\|e_n\| \leq \delta, \; n = 1,2,\ldots \tag{2.1}$$

Relying on the a priori error estimate (1.8) we are then led to try to choose the time steps k_n and the space steps h_n so that for $n = 1,2,\ldots,$

$$L_n \max(C_0 h_n^r |u|_{r,I_n}, \min(C_1 k_n \|\Delta u\|_{I_n}, C_2 k_n^2 \|\Delta u_t\|_{I_n}, C_3 k_n^3 \|\Delta u_{tt}\|_{I_n})) \sim \frac{\delta}{2}. \tag{2.2}$$

Of course, here the quantities $|u|_{r,I_n}$, $\|\Delta u\|_{I_n}$, etc. are not known in advance. However, it is possible to estimate these quantities through the computed solution U and this leads to the adaptive algorithm which we shall now describe. Let us first introduce the discrete counterparts $\Delta_n : H_0^1(\Omega) \to S_n$ of the Laplace operator Δ defined as follows:

$$(-\Delta_n \varphi, \psi) = (\nabla\varphi, \nabla\psi), \quad \forall \psi \in S_n. \tag{2.3}$$

Let us now for simplicity assume that $r = 2$ and let us recall that by elliptic regularity

$$|\varphi|_2 \leq C\|\Delta\varphi\|$$

if $\varphi = 0$ on Γ. This means that in (2.2) the quantity $|u|_{2,I_n}$ may be replaced by $C\|\Delta u\|_{I_n}$. Replacing now in (2.2) Δ by Δ_n, u by U and time derivatives by simple difference quotients, we are led to the following criterion for choosing the local time and space steps:

$$L_n \max(C_0 h_n^2 d_{1n}, \min(C_1 k_n d_{1n}, C_2 k_n^2 d_{2n}, C_3 k_n^3 d_{3n})) \sim \frac{\delta}{2}, \quad (2.4)$$

where

$$d_{1n} = \|\Delta_n U_n^-\|, \quad (2.5a)$$

$$d_{2n} = \left\|\frac{\Delta_n U_n^- - \Delta_{n-1} U_{n-1}^-}{k_n}\right\|, \qquad n > 1, \quad (2.5b)$$

$$d_{3n} = \frac{1}{k_n}\left\|\frac{\Delta_n U_n^- - \Delta_{n-1} U_{n-1}^-}{k_n} - \frac{\Delta_{n-1} U_{n-1}^- - \Delta_{n-2} U_{n-2}^-}{k_{n-1}}\right\|, \quad n > 2, \quad (2.5c)$$

and where we set $d_{21} = d_{31} = d_{32} = \infty$. To determine h_n and k_n from (2.4) we would in principle have to (approximately) solve nonlinear equations since the d_{in} depend on h_n and k_n. In our implementation however we have simply used the predicted values of h_n and k_n resulting from replacing in (2.4) the d_{in} by the quantities $d_{i,n-1}$ available if U has been computed up to time t_{n-1}. We have also replaced the logarithmic factor L_n by 1. Thus our algorithm for automatic choice of space and time step in (1.4) with $q = 1$ is as follows: For $n = 1,2,\ldots,$ choose

$$h_n = \left(\frac{\delta}{2C_0 d_{1,n-1}}\right)^{1/2} \quad (2.6a)$$

$$k_n = \max\left(\frac{\delta}{2C_1 d_{1,n-1}}, \left(\frac{\delta}{2C_2 d_{2,n-1}}\right)^{1/2}, \left(\frac{\delta}{2C_3 d_{3,n-1}}\right)^{1/3}\right). \quad (2.6b)$$

If the predicted steps according to (2.6) and the corresponding solution U_n^- satisfy (2.4) the steps h_n and k_n are accepted and the computation proceeds, otherwise h_n and k_n are modified accordingly until (2.4) is met.

In our numerical tests with $f \equiv 0$ and $u|_\Gamma \equiv 0$ for $t > 0$, requiring (2.4) to be satisfied up to a factor two, the predictions (2.6) were always accepted.

The adaptive method for the backward Euler method (1.5) corresponding to (2.4) reads:

$$L_n \max(C_0 h_n^2 d_{1n},\ C_1 k_n d_{1n}) \sim \frac{\delta}{2}\ . \qquad (2.7)$$

In [7] we proved under certain natural assumptions an a posteriori error estimate for the backward Euler method with discretization only in time of essentially the following form:

$$\|e_N^-\| \leq L_N \max_{n \leq N} C_1 k_n d_{1n}\ . \qquad (2.8)$$

This estimate clearly justifies time step control for the backward Euler method according to (2.7) and we see that if the computational criterion (2.7) is satisfied, then by (2.8) the time discretization error is controlled globally in time to the given tolerance δ. Note that no previous result of this nature for stiff initial value problems seems to be available in the literature.

Now, it is possible to prove under similar assumptions an a posteriori error estimate for (1.4) with $q = 1$ corresponding to (2.7), that is an a posteriori estimate of essentially the form

$$\|e_N^-\| \leq L_N \max_{n \leq N}(C_0 h_n^2 d_{1n} + \min(C_1 k_n d_{1n}, C_2 k_n^2 d_{2n}, C_3 k_n^3 d_{3n}))\ . \qquad (2.9)$$

By this estimate it follows that mesh control through (2.4) will guarantee that the error is controlled globally to the tolerance δ. The detailed proof of (2.9), which is analogous to the proof of (2.8) given in [7], will appear in a subsequent note.

Remark. Let $\{\chi_1, \ldots, \chi_M\}$ be a finite element basis for S_n and let $A_n = (a_{ij}^n)$ and $B_n = (B_{ij}^n)$ be the corresponding stiffness and mass matrices with elements

$$a_{ij}^n = (\nabla\chi_i, \nabla\chi_j),$$

$$b_{ij}^n = (\chi_i, \chi_j).$$

For $\varphi \in S_n$ with

$$\varphi = \sum_{i=1}^{M} \xi_i \chi_i, \qquad \xi_i \in \mathbb{R},$$

we then have

$$\Delta_n \varphi = \sum_{i=1}^{M} \eta_i \chi_i, \quad \eta_i \in \mathbb{R},$$

where

$$\eta = M^{-1} A \xi, \qquad \eta = (\eta_i), \qquad \xi = (\xi_i). \qquad \square$$

3. Numerical results

In this section we present the results of some numerical experiments using the method (1.6) with mesh control according to (2.6) in the case of the one-dimensional problem

$$\begin{aligned} u_t - u_{xx} &= 0, & 0 < x < 1, \quad t > 0, \\ u(0,t) = u(1,t) &= 0, & t > 0, \\ u(x,0) &= u_0(x), & 0 < x < 1, \end{aligned} \tag{3.1}$$

with initial functions u_0 of varying degree of smoothness. The space meshes were restricted to be uniform subdivisions $\Omega_n = \{J\}$ of $\Omega = (0,1)$ into intervals J of length $h_n = 2^{-m}$, $m \in \mathbb{Z}^+$ with

$$S_n = \{v \in H_0^1(\Omega) : v|_J \text{ is linear } \forall J \in \Omega_n\}.$$

Case I $u_0(x) = \sin(\pi x)$

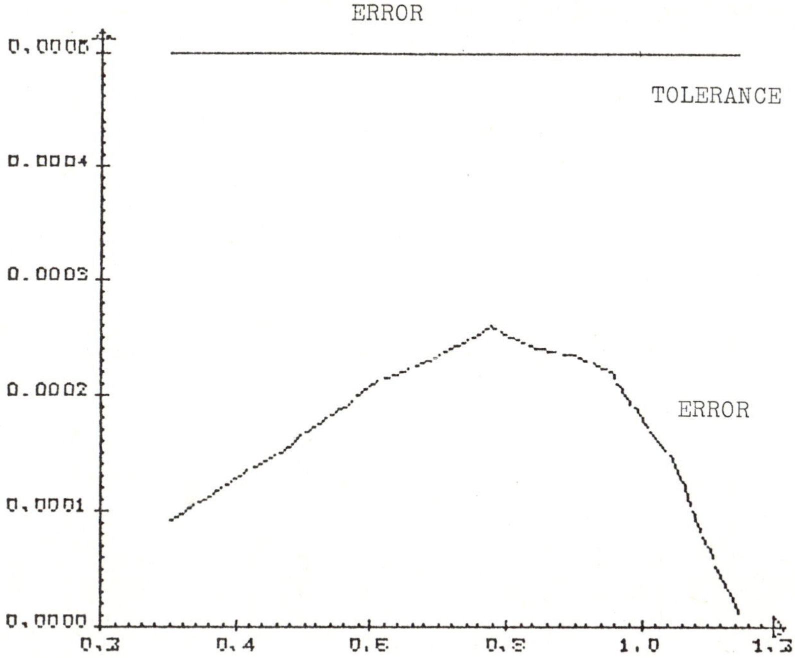

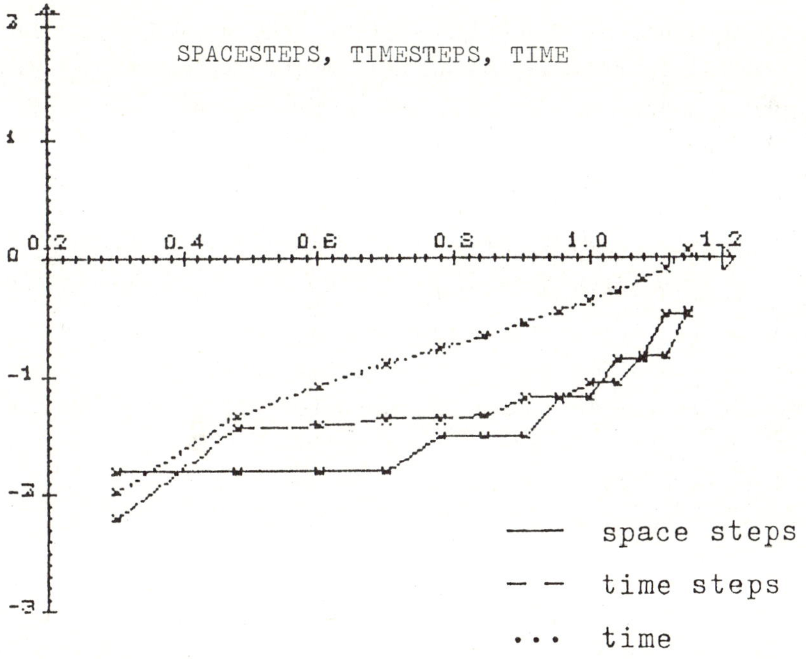

In the computations we used the mesh control (2.6) with (2.6a) replaced by

$$h_n = 2^{-m},$$

$$m = \text{integer part of } {}^2\log\left(\frac{\delta}{2C_0 d_{1,n-1}}\right)^{1/2} .$$

We further used the following values of the constants C_i:

$$C_0 = \frac{1}{8}\ ,\ C_1 = 1,\ C_2 = \frac{1}{6}\ ,\ C_3 = \frac{1}{72}\ ,$$

which are motivated by simple approximation theoretic considerations (cf. [4]).

The following initial functions with different degree of regularity were considered:

(I) $u_0(x) = \sin(\pi x)$,

(II) $u_0(x) = \min(x, 1-x)$,

(III) $u_0(x) = 1$ (non-compatible with boundary data).

In case I the initial function is infinitely smooth while the cases II and III correspond to initial functions in $H^{\frac{3}{2}-\varepsilon}(\Omega)$ and $H^{\frac{1}{2}-\varepsilon}(\Omega)$, $\varepsilon > 0$, respectively. For $t > 1$ the solutions quickly become stationary. For comparison the exact solutions were computed using Fourier series.

In the diagrams below we give the L_2-norm of the error $e_n^- \equiv u(t_n) - U_n^-$ as a function of ${}^{10}\log(n)$ for $t_n \leq 1$. In all three cases we see the remarkable fact that the error, up to roughly a factor two, is kept on a constant level C for all time, where $C \sim 0.2$. The jumps in the error are related to the changes of mesh size in space where the number of nodes are divided by 2 at each change.

For each case we also give the space steps h_n, the time steps k_n, and the discrete time levels t_n as functions of n in a ${}^{10}\log$ - ${}^{10}\log$ diagram. We may compare the h_n and t_n given

Case II $u_o(x) = \min(x,1-x)$

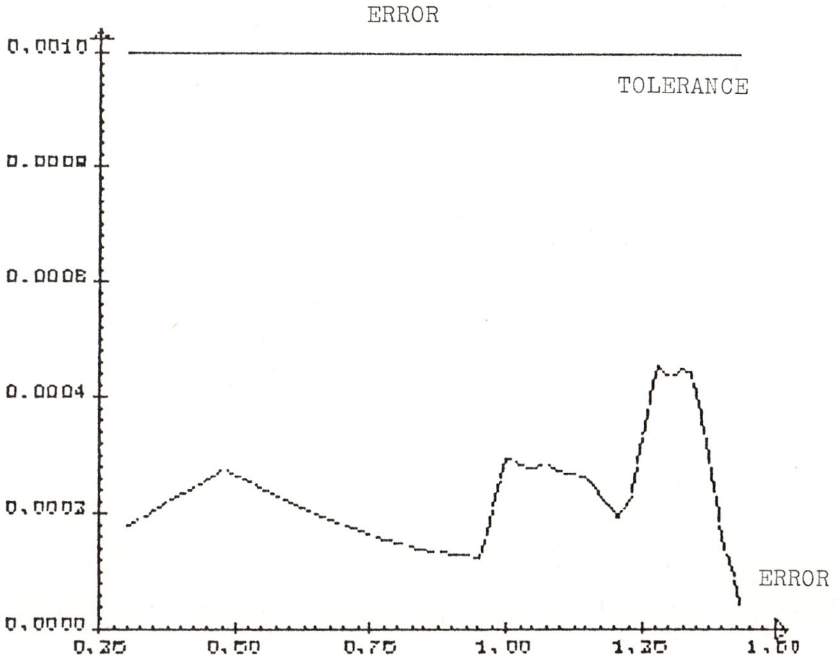

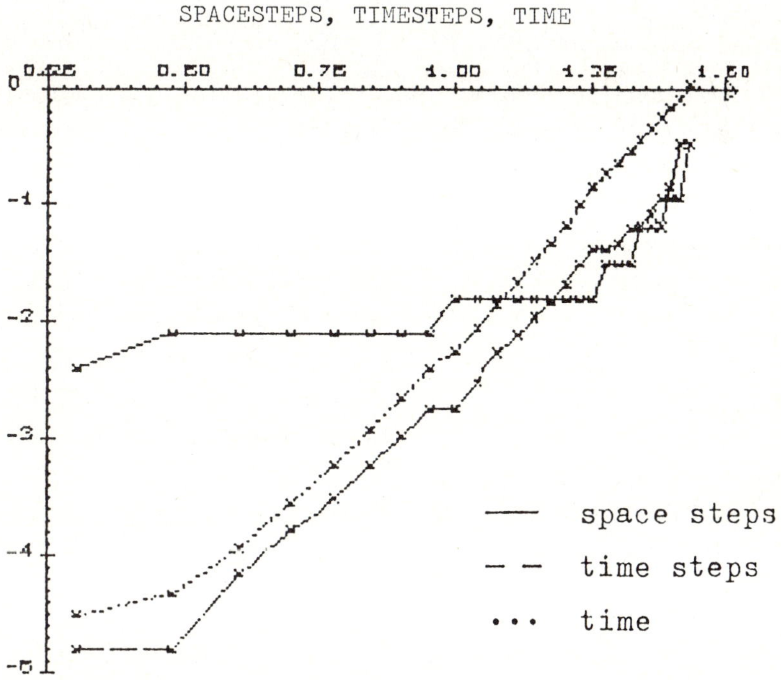

by the computational procedure with predictions of these quantities obtained from (2.2) by estimating analytically the norms $\|\Delta u(t)\|$ etc. In the case II with initial data $u_0(x) = \min(x,1-x)$ we expect to have

$$\|\Delta u(t)\| \sim t^{-\frac{1}{4}}, \quad \|\Delta u_t(t)\| \sim t^{-\frac{5}{4}}, \quad \|\Delta u_{tt}\| \sim t^{-\frac{9}{4}}, \tag{3.2}$$

if t is not very small and $t < 1$, say, and $a \sim b$ indicates that a is (roughly) proportional to b. From (2.2) we would then get

$$k_n \sim t_n^{3/4}, \quad h_n \sim t_n^{1/8}. \tag{3.3}$$

Since

$$n \sim \int_0^{t_n} \frac{1}{k_n}\, dt \sim t_n^{1/4},$$

we thus expect to have in case II

$$t_n \sim n^4, \quad k_n \sim n^3, \quad h_n \sim n^{1/2},$$

i.e., $\log t_n = 4 \log n$ etc. Comparing these predictions with the actual outcome as given in the corresponding log-log diagram we find good agreement. We find a similar good agreement in the other case of non-smooth initial data, i.e., case III.

Case III $u_o(x) = 1$ (non-compatible with boundary data)

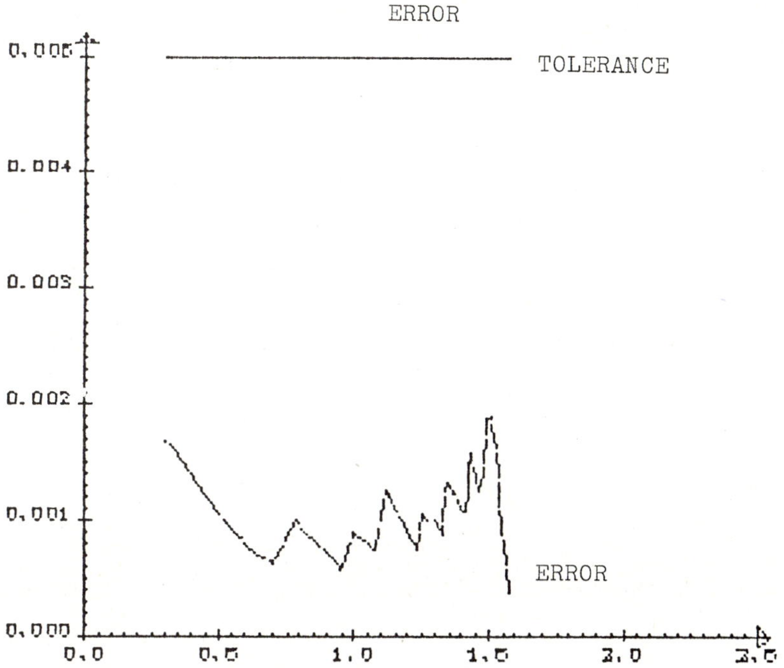

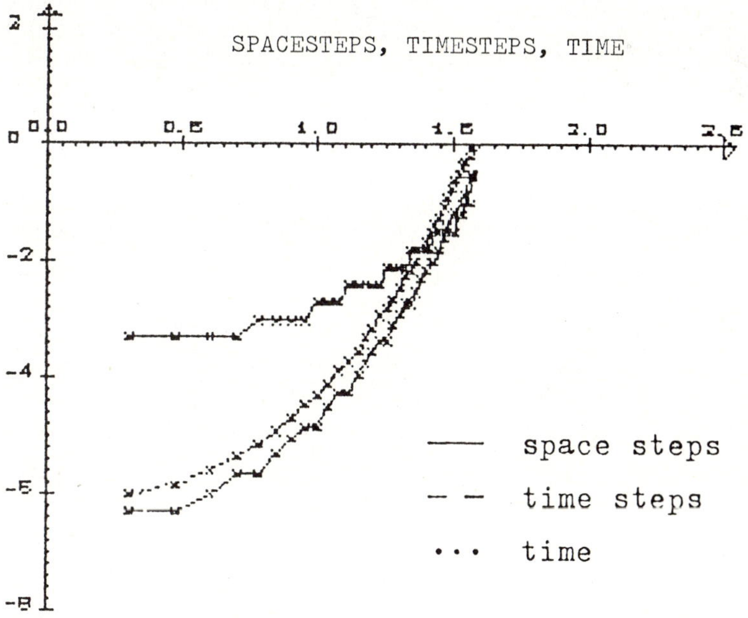

REFERENCES

[1] Ciarlet, P., The Finite Element Method for Elliptic Problem, North Holland, Amsterdam (1978).

[2] Eriksson, K., Johnson, C. and Thomée V., Time discretization of parabolic problems by the Discontinuous Galerkin method, RAIRO, MAN vol. 19 (1985), p. 611-643.

[3] Eriksson, K. and Johnson, C., Error estimates and automatic time step control for non-linear parabolic problems, I, Preprint no. 1985-20, Department of Mathematics. Chalmers University of Technology, Göteborg, to appear in SIAM J. of Numer. Anal.

[4] Eriksson, K., Johnson, C. and Lennblad, J., Optimal error estimates and adaptive time and space step control for linear parabolic problems, Preprint no. 1986-06, Department of Mathematics, Chalmers University of Technology, Göteborg.

[5] Eriksson, K. and Johnson, C., An adaptive finite element method for linear elliptic problems, Preprint no. 1985-13, Department of Mathematics, Chalmers University of Technology, Göteborg.

[6] Johnson, C., Error estimates and automatic time step control for numerical methods for stiff ordinary differential equations, Preprint no. 1984-27, Department of Mathematics, Chalmers University of Technology, Göteborg, submitted to SIAM J. of Numer. Anal.

[7] Johnson, C., Nie, Y. and Thomée V., An a posteriori error estimate and automatic time step control for a backward Euler discretization of a parabolic problem, Preprint no. 1985-23, Department of Mathematics, Chalmers University of Technology, Göteborg, submitted to SIAM J. of Numer. Anal.

AN EFFICIENT NUMERICAL SCHEME FOR VORTICAL TRANSONIC FLOWS

Laszlo Fuchs

Department of Gasdynamics,
The Royal Institute of Technology,
S-100 44 Stockholm, Sweden.

SUMMARY

A combined algorithm using potential/Euler solvers, for transonic flow computations, is described. This combined scheme is substantially more efficient than the basic Euler solver. The potential solver is used to provide initial approximation to the Euler solver, to determine the regions where the flow is vortical and where the Euler solver has to be used. Since the potential solver provides a relatively good approximation to the Euler solution, few (one or two) coupled iterative steps are adequate. The coupled scheme, is faster and requires less computer storage than current Euler solvers.

INTRODUCTION

Numerical methods for flows of fluids have gain rather many industrial applications. In some applications, the (non-linear) potential theory is not a good enough approximation of real flows. These types of compressible flows are treated, in general, by solving the Euler equations. The only underlying assumption for these equations, beside continuum, is that the fluid is inviscid. The potential approximation is obtained from the Euler equations by assuming irrotational motion. For external airfoil problems, the flow is irrotational if the inflow velocity profile is irrotational and if there are no shocks (assuming that the fluid is inviscid and that the vorticity due to the circulation around the airfoil can be represented by a singular surface, i.e. a vortex sheet).

Numerical methods for solving the potential equations require less computer capacity. The potential model consist of a single PDE (both in 2-D and 3-D), whereas the Euler equations consist of 2+d equations, where d is the dimensionality of the problem. (One PDE may be replaced by an algebraic equation if one assumes that the total enthalpy is constant). Furthermore, the character of the potential PDE and the Euler system of PDE's are different. The former is (locally) elliptic whenever the flow is subsonic, and hyperbolic at regions of supersonic flows. The Euler equations are hyperbolic always, independent of the flow. As a result of the different types, these two approximations require different boundary conditions with regard to number and type. The iterative methods for solving the two models are also different. The potential equations can be solved very efficiently by Multi-Grid (MG) [1-3,5] and by Approximate-Factorization (AF) methods [4,5]. The convergence rates achieved by MG methods can be very close, even for supercritical cases, to those attained by MG methods for the Laplace equation [5]. In the case of the Euler equations, most iterative methods use a pseudo time-marching techniques (e.g. [11]). These schemes, even in a MG configuration [6,7], do not attain the efficiency level attained by the potential solvers. MG solvers for the steady-state problem, using flux splitting methods [8] have been reported to be faster, but these methods seem to have low order of accuracy (i.e. large artificial viscosity). All these factors together make Euler solvers to be an 'expensive' alternative to the potential solvers. For these reasons there have been made different attempts to improve the potential model, while keeping its numerical efficiency.

One of the early attempts of non-isentropic potential flow modelling were made by Fuchs [1], who used the transonic small perturbation equation, and included in it a correcting term, that accounted for the increase in entropy behind the shock. Methods for correcting the full potential equations have been suggested by Klopfer and Nixon [9] who included in the potential model the conservation of mass and momentum or momentum and energy across shocks. Hafez and Lovell [10] solve, beside the equation for the potential an additional equation, with a similar structure, for the streamfunction. These models are, however, more limited in applications than the complete set of Euler equations. In this paper we use the potential solution to improve the efficiency of the Euler solver. This is done by providing a good initial approximation to the Euler solver and by allowing a significant reduction of the computational domain. In most applications one does not have to iterate between the potential and the Euler solvers, and a single step of a local Euler correction yields very good results. The combined scheme results in a considerable improvement of numerical efficiency: shorter computational times and reduction in computer memory requirements.

In the following we describe shortly the Euler scheme that we use and how it is modified to accommodate the potential results. Finally, we give some computational examples on the effects of the different modifications on the numerical efficiency.

TRANSONIC FLOW MODELLING

Let p, ρ, u, v, E and H denote the pressure, density, velocity components in x- and y-directions, total energy and total enthalpy. For a perfect gas, with a specific heat ratio γ, the following relations hold:

$$E = p/(\gamma-1)\rho + 1/2\ (u^2 + v^2),$$

and (1)

$$H = E + p/\rho.$$

The Euler equations for two dimensional inviscid flow can be written as:

$$\mathbf{W}_t + \mathbf{F}_x + \mathbf{G}_y = 0 \qquad (2)$$

where $\mathbf{W} = (\rho, \rho u, \rho v, \rho E)^T$

$\mathbf{F} = (\rho u, \rho u^2+p, \rho uv, \rho uH)^T$

$\mathbf{G} = (\rho v, \rho vu, \rho v^2+p, \rho vH)^T$

These equations express the conservations of mass momentum and energy, respectively. The system (2) is hyperbolic for all Mach numbers. The free stream Mach number determines, however, the number of conditions that should be given. This number is determined by the number of incoming characteristics. For inflow boundary, there are 3 or 4 such characteristics, depending on the inflow Mach number (subsonic and supersonic, respectively). On outflow boundaries, one may specify one condition if the flow is subsonic locally, or none if the local Mach number is greater than 1. On solid boundaries one imposes zero normal mass flux. This condition is complemented by a condition on the pressure by computing the normal pressure derivative and extrapolating the pressure from the flow field to the boundary using the normal derivative of the pressure [11].

If the flow is such that the total enthalpy is conserved, then the number of equations (and unknowns) can be reduced by using relations (1). If one further assumes that the changes in entropy are small (due to relatively weak shocks), then by Crocco's relation one readily finds that the flow is also irrotational. For such flows, one may introduce a velocity potential, Φ, and the governing equations simplify to the mass conservation equation:

$$(\rho\, \Phi_x)_x + (\rho\, \Phi_y)_y = 0$$

and (3)

$$\rho^{(\gamma-1)} = 1 + \frac{\gamma-1}{2} M_\infty\, [1-(u^2 + v^2)]$$

with $(u,v)=\nabla\Phi$.

The boundary conditions for (3) are given by specifying Φ on the far-field boundaries and imposing that the mass flux through solid boundaries vanishes.

By introducing the density in the differential equation and rewritting it in a quasi-linear form, one can note that the equation is elliptic or hyperbolic, depending on the local Mach number (smaller than 1 or greater than 1). The far-field conditions has to be modified if the flow is supersonic: the condition on Φ should be supplemented by a condition on its normal derivative on inflow boundaries, and no conditions can be specified on outflow boundaries.

NUMERICAL SCHEMES

Both models have been applied for solving the flow past airfoils. A body fitted mesh is generated, such that the spacing near the airfoil, and especially close to the leading and trailing edges is finer. The potential equation is written in terms of the of the transformed coordinates (in conservative form). The equations are discretized by centered finite differences, and artificial viscosity is added to the density in supersonic flow regions (see [1,4,5]). The discrete equations are solved by an Approximate Factorization (AF)scheme [4,5].

Our Euler solver is based on the code of A. Rizzi [12]. This code has been modified to accommodate MG processing, local mesh refinements and includes the new additional features of the combination with the potential solver. The Euler equations are integrated on each computational cell:

$$\frac{\partial}{\partial t} \iint \mathbf{W}\, dx\, dy + \int (\mathbf{F}\, dy - \mathbf{G}\, dx) = 0. \tag{4}$$

The fluxes in (4) are approximated by averaging [6,7,11-13]. To these discrete equations we add second and fourth order damping terms (see e.g. [11-13]). The final space discretized equations have the form:

$$\frac{d}{dt} (V\mathbf{W}) + (Q+D)\mathbf{W} = 0 \tag{5}$$

where V denotes the computational cell volume, Q is the flux averaged part and D the diffussive (artificial viscosity) part of the space discretization operator. For convenience we denote $P = Q+D$. The time evolution problem is solved by a 3-step Runge-Kutta method. If $\mathbf{W}^n$ denotes the vector of the unknowns at time step n, and we want to advance to time step n+1 the following algorithm is followed:

$$\begin{aligned}
\mathbf{W}^{(0)} &= \mathbf{W}^{n}. \\
\mathbf{W}^{(1)} &= \mathbf{W}^{(0)} - \Delta t\, P\mathbf{W}^{(0)} \\
\mathbf{W}^{(2)} &= \mathbf{W}^{(0)} - {}^{1}/{}_{2}\Delta t\, (P\mathbf{W}^{(0)} - P\mathbf{W}^{(1)}) \\
\mathbf{W}^{(3)} &= \mathbf{W}^{(0)} - {}^{1}/{}_{2}\Delta t\, (P\mathbf{W}^{(0)} - P\mathbf{W}^{(2)}) \\
\mathbf{W}^{n+1} &= \mathbf{W}^{(3)}.
\end{aligned}$$

This scheme is formally, second order accurate in time, and for the linearized 1-D Burger equation it is stable (for all 'Reynolds numbers') when the Courant number, C, is less or equal 2. That is, for a single equation with P as a scalar:

$$\left|\frac{P\Delta t}{\Delta x}\right| \leq 2.$$

This stability conditions places a rather large limitation on the time step if Δt is uniform. Since in most computations one is interested only in the steady state, we abandon time accurate integration and we use locally determined, time steps. After each Runge-Kutta cycle the boundary conditions are updated. The far-field boundary conditions are computed by either far-field values (for incoming characteristics) or by extrapolation from the flow field (for the outgoing characteristics). This basic Euler scheme is rather slow, and it takes between several hundreds to some thousands time steps, to obtain a converged solution.

By improved numerical efficiency we mean shorter computational times and less computer memory for the solution of a given problem. Improved efficiency may be attained by one or a combination of the following measures:

* Modified modelling (reduction in number of unknowns).
* Improved discretization techniques, so that the number of degrees of freedom can be reduced without hamperring accuracy (e.g. adaptive mesh refinements and higher order schemes).
* Faster iterative methods.

In the present work we address the first possibility. The other options have been studied and are being at different stages of testings.

To improve the efficiency of the Euler solver we use the potential solver. The potential equation is solved initially. The potential results are interpolated (averaged) to the cell centers, where they are defined in the current finite-volume scheme. The potential equation provides the velocity vector, the density and the pressure. The last two quantities are computed by using isentropic relations. The potential solution is used by the Euler solver in two ways: It is used as an initial approximation for the Runge-Kutta steps and it provides boundary conditions when the Euler computational domain is restricted to regions where the potential approxmiantion is not good enough. Once the Euler equations, in the reduced domain, are solved, one has to update the potential solver (i.e. correct for the rotational part of the flow field). This can be done by adding a correction to the isentropic density formula (3). Our solution algorithm has the following form:

Step i. Solve the potential equation for Φ and ρ_p:

$$\nabla\cdot(\rho_p \nabla\Phi^{(1)}) = 0$$

set n=1. Define the 'Euler domain' to include at least the regions downstream of the shocks.

Step ii. Interpolate the correction due to the potential solution (i.e. the differences between the current potential and the current Euler solution) to the Euler mesh.

Step iii. Solve the Euler problem ($P\mathbf{W}=\mathbf{0}$) in the 'Euler domain' (possibly, a subdomain of the full potential domain), to certain accuracy, using the modified initial approximation (of step ii.). That is, solve approximatly the steady problem

$$P\mathbf{W}^{(n)} = \mathbf{0}.$$

Step iv. Compute the corrections for the potential solver:

$$\Delta\rho_p = \rho_E^{(n)} - f(\mathbf{u}_E)$$

where f is the value of the density using the isentropic relation (3) and the Euler velocity vector, $\mathbf{u}_E$. Note that the right hand side of the potential problem is also modified.

Step v. Solve the modified potential equation:

$$\nabla\cdot(\bar{\rho}_p \nabla\Phi^{(n+1)}) = R^{(n+1)}$$

where $\bar{\rho}_p = f(u_p,v_p) + \Delta\rho_p$

and $R^{(n+1)} = \nabla\cdot(\bar{\rho}_p\nabla\Phi^{(n)} - \rho_E\mathbf{U}_E^{(n)})$.

Step vi. If the changes in the density are too large set $n\leftarrow n+1$ and go to Step ii.

The scheme described above can be regarded as a defect correction scheme for the potential equation (with the defects correcting for the entropy rise due to the rotational character of the flow). The better the potential model in approximating the Euler equations is, the less number of defect correction steps one has to make. The number of steps, that one should make can be estimated in terms of the deviation from isentropic flow (i.e. in terms of entropy production). It is well known that the entropy production across a shock is proportional the shock strength to the third power. Thus, for transonic flows, the iterative process above, the number of defect correction cycles can be as low as 1 or 2. That is, in most cases it is enough if both the potential and the Euler equations are solved once or at most twice. The amount of additional computational work in solving the potential equation is only a small fraction compared to that required for the solution of the full Euler problem. This is so since the potential problem consists of a single PDE and the iterative process has higher rate of convergence than that of the corresponding Euler solver. It should be noted that when the above iterative scheme converges, the solutions (the density) of the potential and the Euler equations are identical.

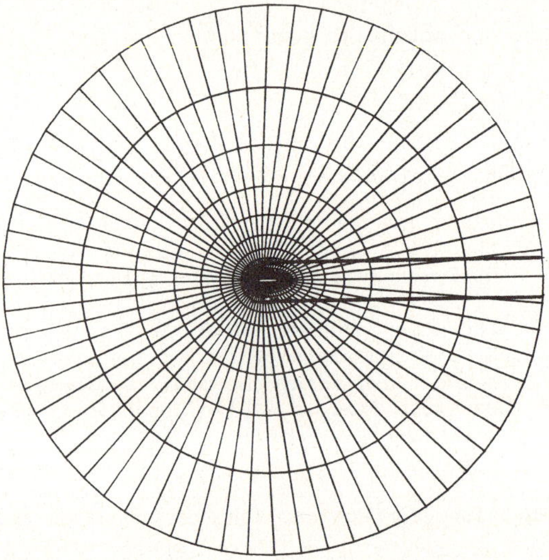

Figure 1.a: The computational grid for the airfoil problem. The potential problem (3) is solved in the whole domain, while the Euler equations (2) are solved only in a limited (typically a rectangular domain as shown in the figure for case a).

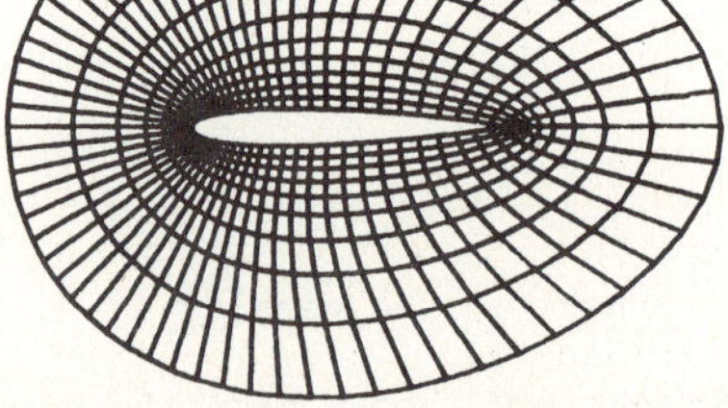

Figure 1.b: Enlargement of the computational mesh near the airfoil.

COMPUTATIONAL EXAMPLES

The algorithms that are described above, have been applied for the solution of the flow past an airfoil (NACA 0012). This problem has been solved by many different schemes and therefore is a good candidate for studying our scheme.

The mesh that was used is shown in Figures 1. For subsonic flows the potential solver provides an exact approximation to the Euler solver. The only errors in the initial approximation are due to interpolating the potential solution from the potential to the Euler grid. For supercritical cases, the errors in the initial approximation are larger, and depend on the strength of the shock (i.e. freestream Mach number, M_∞, and the angle of attack, α).

First, we have used the original potential and the Euler solvers for two supercritical cases; a: $M_\infty = 0.8$, and $\alpha = 0°$, and b: $M_\infty = 0.85$ and $\alpha = 1°$. The potential and the Euler solutions differ considerably with increasing shock strength (see Fig. 2.a and 2.b, respectively). The convergence factor of the unmodified Euler solver for cases a and b are 0.980 and 0.984, respectively. The convergence factors are defined as the mean reduction of the error for a computational effort equivalent to one (three staged) time step in the Euler solver on the full (finest) grid. The slow asymptotic rates of convergences are not improved by using different initial approximations (e.g. by coarse grid approximation or by potential solution). However, the use of good initial approximations does improve the initial rate of convergence (to a level better than that due to discretization errors). When the initial approximation is computed by interpolating a coarse grid solution, the number of time steps may be halved (to about 450). When we use the interpolated potential solution as initial approximation, the Euler equations converge well, in 110 iteration, giving a mean convergence factor of 0.958 (Fig 3.a) and 0.962 (Fig. 4.a) for cases a and b respectively. When the Euler equations are solved only in a limited, 'Euler region' (see Fig. 1) the convergence rate improves further. The convergence factors for the two cases, are 0.888 (Fig. 3.b) and 0.929 (Fig. 4.b), respectively. It should be noted that in these computations the coupled algorithm was used only once (n=1). That is, the local Euler solution was not used to improve the outer (potential) solution. Therefore, the extent of the Euler region in case b had to be extented compared to case a, leading to a reduction in convergence rate. For comparison, one may consider one of the best MG-Euler solver as reported by Jameson [13]. That scheme uses a four stage scheme (and therefore the 'work unit' in that case is larger by about 33% compared to the current three stage scheme). The convergence factors reported in [13] are somewhat larger than 0.9. That is, the current scheme can provide a solution, using much simpler schemes, at less programming and computational effort (and requires less computational memory, since the Euler equations are solved only on parts of the domain).

The limited 'Euler domain' solutions (as shown in Fig. 3.b and 3.c) have been compared to the full region Euler solutions. In all the computed cases, the differences could be made as small as one wishes by enlarging the 'Euler domain'. Compared to the results published in the literature it was found that for practicle applications, it was enough to limit the 'Euler domain' to include the shocks and the regions behind these shocks. Thus, in most cases the computational domain could be reduced by a factor between 2 and 4, without loss of (graphically noticable) accuracy.

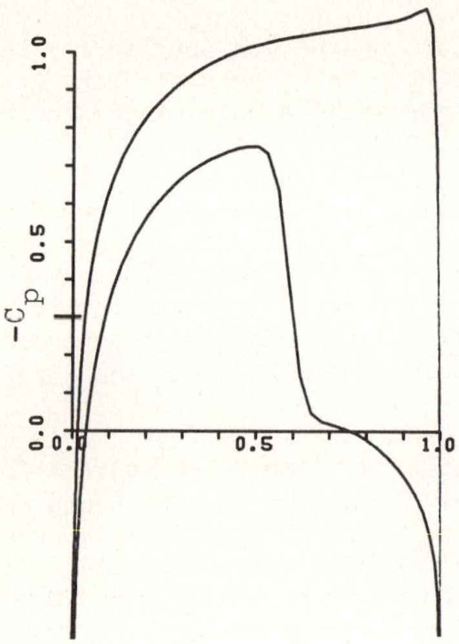

Figure 2.a: The pressure coefficient, C_p, on the airfoil. Full grid potential solution. $M_\infty=0.85$, $\alpha=1°$.

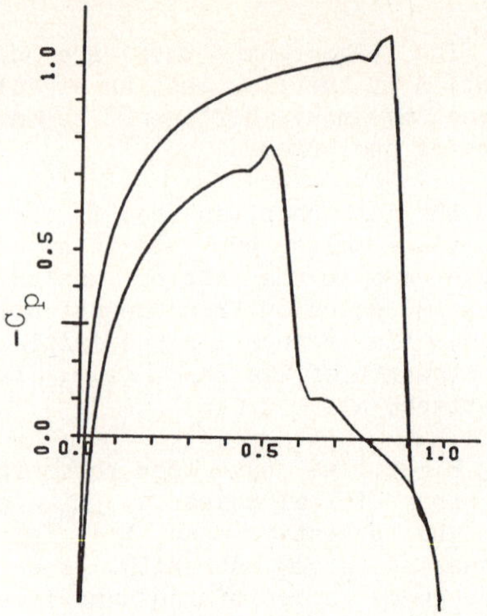

Figure 2.b: The pressure coefficient, C_p, on the airfoil. Full grid Euler solution. $M_\infty=0.85$, $\alpha=1°$. Mean convergence factor, $\theta=0.984$ (not fully converged).

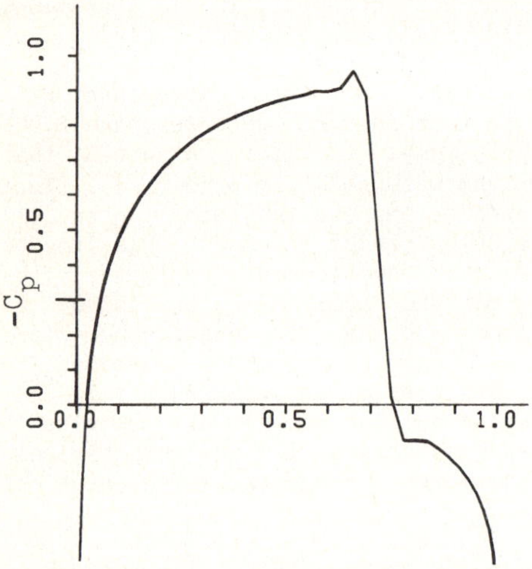

Figure 3.a: C_p on the airfoil. Full Euler solution using interpolated potential solution as initial approximation. $M_\infty=0.85$, $\alpha=0°$. $\theta=0.958$

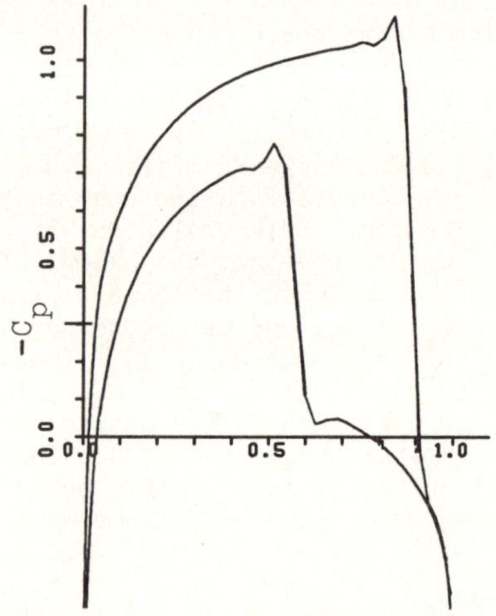

Figure 3.b: C_p on the airfoil. Full Euler solution using interpolated potential solution as initial approximation. $M_\infty=0.85$, $\alpha=1°$. $\theta=0.962$

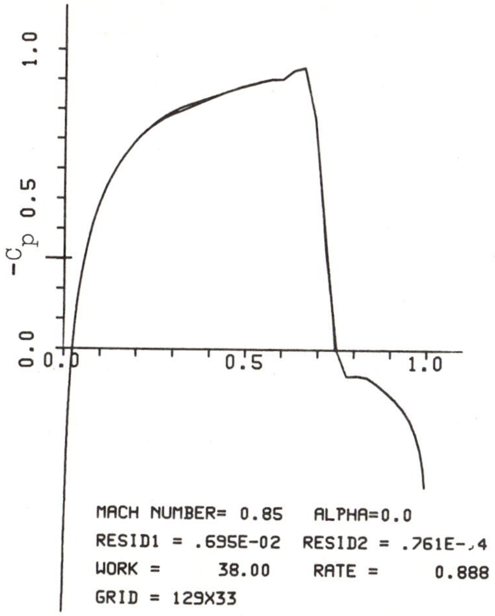

Figure 4.a: C_p on the airfoil. Euler solution only in a limited 'Euler domain'. M_∞=0.85, α=0°. θ=0.888

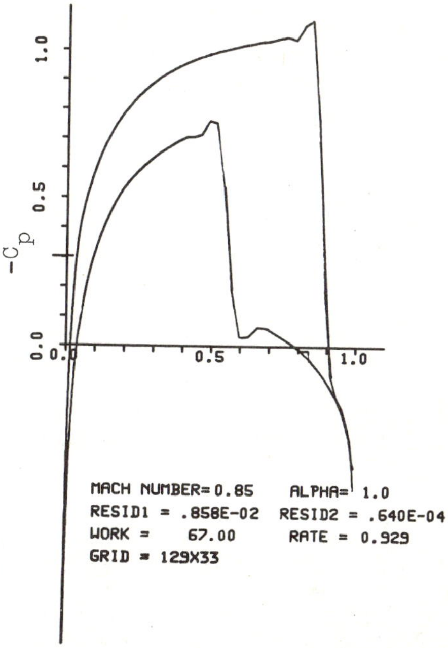

Figure 4.b: C_p on the airfoil. Euler solution only in a limited 'Euler domain'. M_∞=0.85, α=1°. θ=0.929

CONCLUDING REMARKS

An algorithm combining a potential and an Euler solver has been presented. The solution of each solver is used to correct the approximation to the other equation. At convergence both solutions are identical. For most transonic flows a single iterative coupling step is adequate. The resulting scheme is as accurate as the Euler solver, but it is more efficient in the sense that it requires less computational effort and less computer memory.

ACKNOWLEDGMENTS

The author wishes to thank Prof. A. Rizzi and Mr. C-Y. Gu for providing their original codes which enabled faster progress in this work.

REFERENCES

[1] FUCHS, L.: "Finite-difference methods for plane steady inviscid transonic flows", The Royal Institute of Technology Report. TRITA-GAD-2, ISSN 0281-7721 (1977).

[2] FUCHS, L.: "Transonic flow computation by a multi-grid method", GAMM-Workshop on Numerical Methods for the computation of Inviscid Transonic Flows with Shock Waves, Rizzi, A. and Viviand, H. (eds.) Vieweg & Sohn, 1979, pp. 58-65.

[3] JAMESON, A.: "A multi-grid scheme for transonic potential calculations on arbitrary grids", Proc. 4-th AIAA CFD conference, pp. 122-146 (1979).

[4] HOLST, T. L.: "Implicit algortihm for the conservative transonic full potential equation using an arbitrary mesh", AIAA paper 78-1113 (1978).

[5] GU, C-Y., FUCHS, L.: "Numerical computation of transonic airfoil flows", in Num. Meth. Laminar and Turbulent Flow-IV, Taylor, C., Olson, M.D., Gresho, P.M., Habashi, W.G. (eds.), Pineridge Press, Swansea, 1985, pp. 1501-1512. (For details see: GU, Y-C.: Transsonic flow computations. TRITA-GAD-8, ISSN 0281-7721, 1985).

[6] SCHMIDT, W., JAMESON, A.: "Application of multiple-grid methods for transonic flow calculations". in 'Lecture Notes in Mathematics', Hackbusch, W. and Trottenberg, U. (eds.), Springer Verlag, Berlin, 1982, pp. 599-613.

[7] WILLIAM, J., USAB, Jr., MURMAN, E.M.: "Embedded mesh solutions of the Euler equations using a multiple-grid method", in 'Advances in computational transonics' Habashi W.G. (ed.), Pineridge Press, Swansea 1985, pp. 447-472.

[8] MULDER, W. A.: "Computation of the quasi-steady gas flow in a spiral galaxy by means of a multigrid method", Proc. 2nd Copper-Mountain Multigrid Conference, 1985.

[9] KLOPFER, Ge Euler equationsNon-isentropic potential formulation for transonic flows", AIAA paper 83-375 (1983).

[10] HAFEZ, M., Lovell, D.: "Entropy and vorticity corrections for transonic flows", AIAA paper 83-1926 (1983).

[11] JAMESON, A., SCHMIDT, W., TURKEL, E.: "Numerical solution of the Euler equations by finite volume methods using Runge-Kutta time-marching schemes", AIAA paper 81-1259 (1981).

[12] RIZZI, A. W.: "Damped Euler-equation method to compute transonic flow around wing-body combinations", AIAA J. 10, (1982), pp. 1321-1328.

[13] JAMESON, A.: A multi-grid solution method for the Euler equations, Proc. ICFD conference, Morton K.W. and Bains, M.J. (eds.), Oxford University Press, to appear 1986.

ON THE NUMERICAL TREATMENT OF VISCOUS FLOWS PAST BODIES WITH CORNERS AND EDGES BY BOUNDARY ELEMENT AND MULTIGRID METHODS

F.K. Hebeker
Universität-GHS, Fb.17
D-4790 Paderborn, West Germany

SUMMARY

The slow viscous flow past a spatial body with corners and edges is investigated mathematically and numerically by means of a boundary element method. For the resulting algebraic system a new multigrid solver is designed and analyzed. It proves to be preferable to that introduced earlier by other authors for related problems. A numerical example illustrates the proposed methods, in particular it shows that the theoretical assertions are sharp in a sense.

INTRODUCTION

Recently, boundary element methods have found increasing interest for the numerical treatment of a large variety of problems arising in the applied sciences. But, while the study of these methods in case of smooth boundary surfaces is well advanced now, the extension to nonsmooth surfaces which poses difficult problems (at least from the mathematical point of view) has been investigated only rarely for the threedimensional case. In the present paper we report on some recent extensions of the classical method of Radon [11] and Wendland [13] to the viscous flow problem. For details see the author's paper [6].

Let us consider the homogeneous exterior Stokes problem

$$- \Delta u + \nabla p = 0, \quad \operatorname{div} u = 0 \quad \text{in } \Omega',$$

$$u|_{\partial\Omega} = g, \qquad u|_{\infty} = 0 \quad . \tag{1}$$

Here denotes $v = u_\infty + u$ the vector velocity field (u_∞ the uniform onflow at infinity), and p the scalar pressure field of the flow (at constant viscosity $\nu = 1$, say). In the body-fixed frame is $\Omega' = \mathbb{R}^3 \setminus \overline{\Omega}$ the flow region, exterior to a rigid body $\overline{\Omega}$ which eventually has corners and edges. Nonconservative exterior forces are excluded here, for a new approach to this ("boundary element spectral method") we refer to Borchers-Hebeker-Rautmann [1].

THE BOUNDARY INTEGRAL EQUATIONS

A hydrodynamical potential theory, fundamental to any kind of boundary element method, is available for (1) in the smooth-boundary case [9]. In particular we have (hydrodynamical) potentials of the simple layer $V\phi$ and of the double layer $W\phi$ (where the surface source ϕ is a vector field on the boundary $\partial\phi$) which solve the differential equations of (1) for any ϕ.

In the nonsmooth-boundary case the Gaussian integral for a constant surface-source $c \in \mathbb{R}^3$ is

$$W c(x) = C_\Omega(x)\, c \quad . \tag{2}$$

Here the characteristic matrix $C_\Omega(x)$ appears, defined by

$$C_\Omega^{ij}(x) = \int_{S_\Omega(x)} \Theta_i \Theta_j \, do_\Theta \quad ,$$

where $S_\Omega(x)$ is as follows: if $x \in \partial\Omega$, then the half-tangents of x to Ω form the mantle of a tangent pyramide (vertex at x) which cuts out of the unit sphere $S_2(x)$ (center at x) an area $S_\Omega(x)$. Consequently, put $S_\Omega(x) = \emptyset$ if $x \in \Omega'$ and $S_\Omega(x) = S_2(x)$ if $x \in \Omega$. Further, $\Theta = z - x$ $(z \in S_2(x))$ are the polar coordinates on the unit sphere. Note that $C_\Omega(x) = 1/2I$ (unit matrix) if x is a "smooth" point of $\partial\Omega$. On grounds of (2) we can calculate the jump relations of $W\phi$ which lead to our boundary integral equation.

We are looking for a solution of (1) in terms of

$$u = (W + \eta V)\psi \;, \quad p \text{ analogous.} \tag{3}$$

Here η is a constant free parameter. By adapting this ansatz to the given boundary data we obtain the system

$$(C_\Omega - W - \eta V)\psi = -g \quad \text{on } \partial\Omega \tag{4}$$

of boundary integral equations. It differs much from that of the smooth-boundary case, since C_Ω may degenerate and W is (strongly) singular in general.

By extending the Radon-Wendland method we obtain the following result [6] : assume that the boundary is piecewise smooth and has "convex corners" only. If

$$\sup_{x\in\partial\Omega} \| C_\Omega(x) - \frac{1}{2} I \| \leq \omega < \frac{1}{2} \tag{5}$$

(spectral norm of the matrix) holds, then (4) is a Fredholm system. If additionally η is chosen positive, then (4) has a unique solution $\psi \in C(\partial\Omega)$ for any boundary data $g \in C(\partial\Omega)$. The corresponding potential (3) is the unique solution of the Stokes problem (1).

In many practical situations we have " convex corners" only, but the condition (5) is difficult to survey: in the particular case of conical corners it is equivalent to the restriction that Ω must not have spines (cf. [13] for the Laplace equation).

An important feature of (4) is that no regularity property seems to be known up to now. This of course has important consequences for the numerical analysis as will be shown now.

THE BOUNDARY ELEMENT METHOD

An efficient boundary element method to solve (4) has been developed in [5] for the smooth-boundary case. This approach will be extended here to the general case following the lines of Bruhn-Wendland [2]. We assume that the boundary $\partial\Omega$ is represented in a satisfactory way by means of a global coordinate frame, for instance, we assume

$$x \in \partial\Omega \; ; \quad x = F(\Theta,\varphi) \; , \quad (\Theta,\varphi) \in [0,1]^2 \qquad (6)$$

(normalized polar coordinates). A large class of bodies is represented in this way, but the restriction is by no means essential for the following developments.

The approximation method then consists of a decomposition of the parameter space $[0,1]^2$ of the boundary into small quadratic elements of mesh-size $h > 0$. We also assume that all the corners and edges of the body are located on the boundary lines of these elements. As trial functions we use globally continuous and piecewise bilinear polynomials on the parameter space, subordinated to the given quadrangulation. For practical applications of course one should resort to a decomposition of the parameter space into triangular elements [10] and eventually to adaptive mesh generation methods.

Then we are looking for an approximant ψ_h in this trial space by solving that discrete system obtained from (4) by means of a suitable collocation procedure, which may formally be written as

$$P_h(C_\Omega - W - \eta V)\psi_h = -P_h g . \qquad (7)$$

This is equivalent to a linear algebraic system, consisting of a nonsparse and nonsymmetric but relatively small and well conditioned system matrix.

Following [2] we can prove uniform convergence of this approximation procedure [6] :

$$\sup_{\partial\Omega} |\psi_h - \psi| \to 0 \qquad \text{when} \quad h \to 0 . \qquad (8)$$

In contrast to the smooth-boundary case we cannot sharpen this asymptotic convergence due to the lack of regularity of the solution ψ of (4). As an important consequence the question of numerical quadrature in (4) is not the urgent one as in the smooth-boundary case. Hence the integrals in (4) may be evaluated by the simple midpoint rule applied on the small quadratic elements. A different important consequence will be discussed in the next section.

A NEW MULTIGRID PROCEDURE

In [5] a multigrid procedure is developed concerning the smooth-boundary case. At the same time Schippers [12] has introduced several multigrid schemes for nonsmooth boundaries. Due to the lack of regularity these schemes prove to be convergent but without any positive convergence order. In this section we propose an improved multigrid procedure where for the spectral radius $\rho_h = O(h \cdot \log 1/h)$ is shown even without any regularity result on (4).

Following [12] let us rewrite the system (4) as

$$(D + C)\psi = -g \qquad \text{on } \partial\Omega , \qquad (9)$$

where D denotes the singular principal part of the Fredholm operator in (4) (from the double-layer potential we take the integral over a small neighborhood of the singularity only, the remainder is included into the compact operator C). Then (7) may be written as

$$(D_h + C_h)\psi_h = -g_h \quad . \tag{10}$$

As a crucial fact D is an invertible and local operator, hence D_h is a "small" sparse matrix which is "quickly" inverted. On the other hand C is compact, hence C_h has the "smoothing property".

But the equivalent second-kind system [12]

$$(I_h + D_h^{-1} C_h)\,\psi_h = -\,D_h^{-1} g_h$$

eventually results in a poor smoothing procedure only, since C_h smooths but D_h^{-1} in general does not preserve this property. Hence a better multigrid procedure [6] stems from this system, equivalent to (10):

$$(I_h + C_h D_h^{-1})\,\chi_h = -\,g_h \ , \tag{11}$$

where the solution χ_h of it is related to ψ_h by $D_h\psi_h = \chi_h$. The latter equation is quickly solved without expense whereas (11) has improved smoothing properties.

For the resulting model two-grid iteration (using a coarse grid of doubled mesh-size) we have proved in [6]: the sequence $(\chi_h^{(k)})$ of approximants converges to the solution χ_h of (11) in the sense of

$$\sup_{\partial\Omega} |\chi_h^{(k)} - \chi_h| \le (\text{const.}\ h\cdot\log\frac{1}{h})^k \cdot \sup_{\partial\Omega} |\chi_h^{(o)} - \chi_h| \ , \tag{12}$$

and the same holds for the corresponding sequence $\psi_h^{(k)} = D_h^{-1}\chi_h^{(k)}$.

Hence the spectral radius of this iterative procedure is $\rho_h = O(h\cdot\log\frac{1}{h})$.

A NUMERICAL EXAMPLE

The boundary element method (7) has been tested against a constructed flow past the body, represented in polar coordinates by

$$r \le 1 + \varepsilon \cdot \sin^3 2\Theta \cdot \cos 3\varphi + \varepsilon \cdot \sin 2\Theta \tag{13}$$

($0 \le \Theta \le \pi$, $0 \le \varphi \le 2\pi$). In case of $\varepsilon = 0$ the body degenerates to a ball. When $\varepsilon > 0$ some conical corners appear at the poles $\Theta = 0$ and $\Theta = \pi$. And when $\varepsilon \to 0.5$ additionably a spine appears at $\Theta = 3/4\,\pi$ (see Fig. 1). Hence from the theory reported above we conclude: when $0 \le \varepsilon < 0.5$ the procedure should converge, when $\varepsilon = 0.5$ a singular behaviour should turn out. This has been confirmed in the following test series: at mesh-size $h = 1/16$, $\eta = 0.8$ and by solving exactly the linear algebraic system by Gauss' algorithm we obtain as a result:

Table 1 Error of the computed flow, depending on the deformation parameter ε .

ε	o.o	o.1	o.2	o.3	o.4	o.5
relative error %	5.5	5.6	6.1	7.5	9.2	16.9 (!)

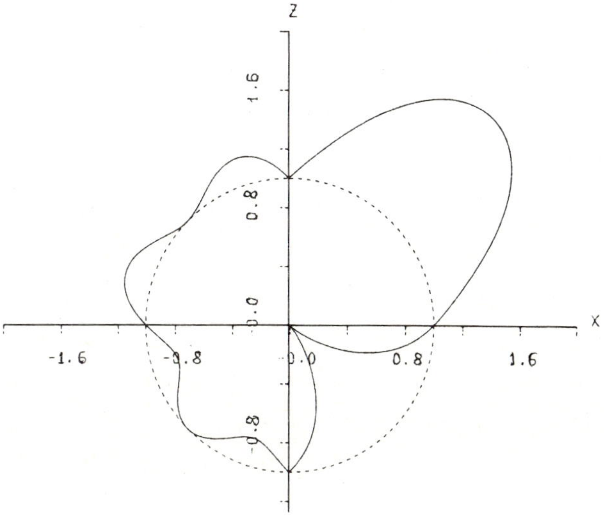

Fig. 1 Irregular body, a spine arising (cross section x,z - plane)

REFERENCES

[1] BORCHERS, W., HEBEKER, F.K., RAUTMANN, R.: "A boundary element spectral method for nonstationary viscous flows in 3 dimensions". E.H.Hirschel (ed.): Finite Approximations in Fluid Mechanics, Vol.14 of Notes on Numerical Fluid Mechanics, Vieweg. Braunschweig/Wiesbaden 1986, 14-28.

[2] BRUHN, G., WENDLAND, W.L.: "Über die näherungsweise Lösung von linearen Funktionalgleichungen". L.Collatz et.al. (eds.): Funktionalanalysis, Approximationstheorie, Numerische Mathematik. Basel 1967, 136-164.

[3] GRISVARD, P.: "Elliptic Problems in Nonsmooth Domains". Boston 1985.

[4] HACKBUSCH, W.: "Multigrid Methods and Applications". Berlin 1985.

[5] HEBEKER, F.K.: "Efficient boundary element methods for 3-D exterior viscous flows". 35 pp. Submitted for publication.

[6] HEBEKER, F.K.: "On the numerical treatment of viscous flows past bodies with corners and edges by boundary element and multigrid methods". 35 pp. Submitted for publication.

[7] KLEINMAN, R.E., WENDLAND, W.L.: "On Neumann's method for the exterior Neumann problem for the Helmholtz equation". J. Math. Anal. Appl. 57 (1977), 170-202.

[8] KRAL, J.: "Integral Operators in Potential Theory". Berlin 1980.

[9] LADYZHENSKAJA , O.A.: "The Mathematical Theory of Viscous Incompressible Flow". New York 1969.

[10] NEDELEC, J.C.: "Approximation des équations intégrales en mechanique et en physique", 127 pp. Ecole Polytechnique, Palaiseau 1977.

[11] RADON, J.: "Über die Randwertaufgaben beim logarithmischen Potential". Akad. Wiss. Wien, Math.-Naturw. Kl., Sitz.-Ber. Abt. II A, vol. 128 (1919), 1123-1167.

[12] SCHIPPERS, H.: "Multigrid methods for boundary integral equations". Numer. Math. 46 (1985), 351-363.

[13] WENDLAND, W.L.: "Die Behandlung von Randwertaufgaben im $\mathbb{R}^3$ mit Hilfe von Einfach- und Doppelschichtpotentialen". Numer. Math. 11 (1968), 380-404.

[14] WENDLAND, W.L.: "Boundary element methods and their asymptotic convergence". P. FILIPPI (ed.): Theoretical Acoustics and Numerical Techniques. Wien 1983, 135-216.

A Technique for Structural Instability Analysis

B. Kröplin
Anwendung Numerischer Methoden
Universität Dortmund
D-4600 Dortmund

Summary

In the paper some methods are discussed, which are in use for static and dynamic stability analysis for thin walled shell structures. A strategy for investigation of secondary solution branches without pinpointing at the bifurcation point is lined out.

1. Introduction

In buckling of very thin walled shell structures snap throughs appear, when the so called critical load is exeeded. The load carrying capacity decreases rapidly. The structure moves from the prebuckling to the postbuckling state and a sequence of rapidly changing buckling patterns is passed. In figure 1 a typical load deflection curve is given. In figure 2 some calculated buckling patterns between the prebuckling state and the postbuckling state are shown.

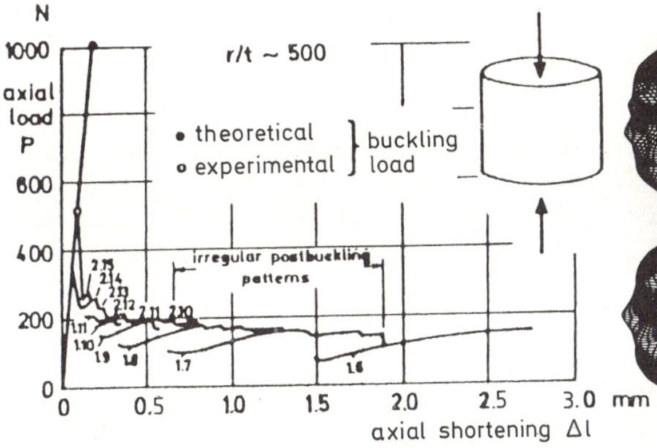

Figure 1: Typical load deflection curve for a thin walled cylindrical shell after /1/.

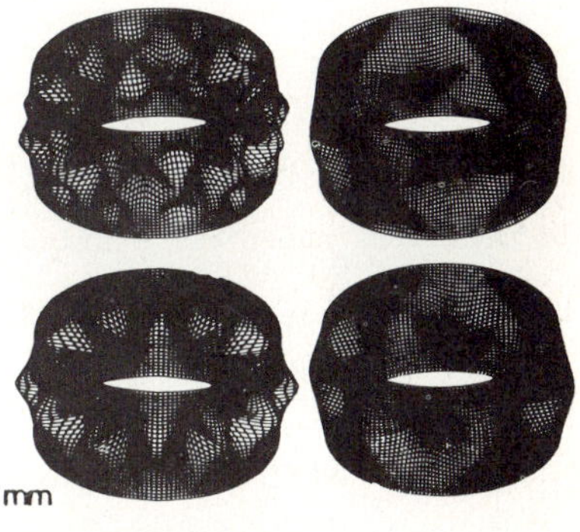

Figure 2: Some buckling patterns in the postbuckling state.

This non linear buckling behavior is well known from experiments /1/. The static postbuckling patterns which are finally reached in the deep postbuckling range are reproducible, while the unstable dynamic behaviour seems to be not.

However also the static buckling patterns are not unique, since also in the experiment different patterns are obtained if transient perturbations are applied during the buckling process. Hence different stable patterns may be related to the same load level.

For a reliable safety analysis of thin walled structures the limit load, the imperfection sensitivity and the lowest stable postbuckling load are of interest, further the grade of stability of the obtained solution.

The above described phenomenon, the rapid movements from the prebuckling to the postbuckling state, is clearly a dynamic one. In spite of this, static strategies have been lined out /2,3/ in order to compute postbuckling pathes and to estimate the load carrying capacity in the postbuckling region. The dynamic analysis avoids numerical difficulties and leads for a non zero damping to a postbuckling pattern, depending on the introduced damping. Different damping may lead to a different postbuckling pattern.
Static calculations have to select the path which is to follow in the non linear solution manifold, where a large number of pathes may exist. Path tracing methods, e.g. the well known constant arc lenght method /2/, are developed in order to follow a solution path in the static unstable region. These methods have to cope with non uniqueness and bifurcations on the followed solution path and path branching. Since pathes may exist, which are very close to each other,but do not intersect, a reliable estimate of the lowest post buckling load can only be given, if all pathes are followed.
However in both static and dynamic methods the stability of the obtained solution which means the sensitivity against a perturbation is not known.

In order to estimate the stability of a solution branch in a static manner a perturbation strategy has been designed. Some results are presented here.

2. Strategy

A number of strategies have been developed for buckling and postbuckling calculations. The strategies have to cope with non uniqueness, bifurcations numerical difficulties in snapping regions.

The most complete form is a dynamic calculation in time domain (1):

$$\underline{\underline{M}}\ddot{\underline{x}} + \underline{\underline{D}}\dot{\underline{x}} + \underline{C}^{s}_{(x)}\,\underline{x} - \underline{P}_{(t)} = 0 \qquad (1)$$

$\underline{M}$: mass matrix

$\underline{D}$: damping matrix

$\underline{C}^{s}_{(x)}$: structural (secant) matrix

$\underline{x}$: vector of unknowns

$\underline{P}_{(t)}$: external force

$\dot{\underline{x}}$: velocity

$\ddot{\underline{x}}$: acceleration.

The structural matrix contains geometrical non linear effects, like large displacements. Since a realistic material damping is not known the damping matrix $\underline{D}$ is used as an artificial device in order to find a static postbuckling pattern. This means also, that different dampings may lead to different solution and different postbuckling states in dynamic analysis.

After discretisation in time domain by means of an implicit operator equation (1) takes the form given in (2). The secant matrix is not known at the beginning of an increment, therefore it is approximated by the tangent matrix. An iteration has to be carried out in order to satisfy equation (2). The time operators and the residual depend on the time integration scheme. A recent survey of time discretisation methods is given in /4/. Dynamic analysis has to satisfy high requirements with respect to stability and accuracy since errors may grow uncontrolled during the calculation. Implicit schemes are known, which are unconditionally stable in the linear case. The accuracy depends on the chosen time step:

$$(\underline{C}^{s}_{(\Delta \underline{x})} + \xi \underline{D} + \eta \underline{M})\ \Delta\underline{x} - \underline{p}^{u} = 0 \qquad (2)$$

$\underline{C}^{s}_{(\Delta x)}$: structural (secant) stiffness matrix in the increment

$\underline{D}$: structural (secant) damping matrix in the increment

$\underline{M}$: mass matrix

ξ : time operator of first order

η : time operator of second order

$\Delta\underline{x}$: increments of the unknowns

$\underline{p}^{u}$: load level and residuum of the previous steps.

While the dynamic problem considers the boundary problem and the initial value problem the static solution deals only with the boundary problem (3). Therefore if the structural secant matrix is nonlinear, a number of solutions may exist for every external load level $\underline{p}$:

$$\underline{C}^{s}_{(\underline{x})}\ \underline{x} - \underline{p} = 0 \qquad (3)$$

$\underline{C}^{s}_{(\underline{x})}$: structural (secant) stiffness

$\underline{x}$: unknown displacements and stress components

$\underline{p}$: external forces.

The solutions may be stable or unstable in the static sense. The solutions are normally obtained by following solution-pathes in small incremental steps. In every step an iteration of the Newton-method is carried out. Since the structural secant matrix may be singular, it is often necessary to stabilize the iteration by special methods. One way to cope with limit points is to apply the above mentioned constant arc lenght-method, which adjusts the load level such, that also in static unstable regions solutions can be reached /2/. Another device to enlarge the size of the increments and apply the load all at once is to introduce an artificial quasi viscous damping. The iteration is then interpreted as time dependent process (4):

$$\underline{D}\dot{\underline{x}} + \underline{C}^{s}_{(\underline{x})}\ \underline{x} - \underline{p} = 0 \qquad (4)$$

$\underline{D}$: artificial damping

$\underline{C}^{s}_{(\underline{x})}$: structural (secant) matrix

$\underline{x}$: unknown displacements and stress components

$\dot{\underline{x}}$: velocities.

With an implicit operator in time domain equation (5) is obtained. (see /4/):

$$(\underline{D}^* + \underline{C}^T_{(\underline{x})})\ \underline{x} - \underline{p}^u = 0 \tag{5}$$

$\underline{D}^* = \underline{p}^u / \Delta \underline{x}$: diagonal artificial damping matrix

$\underline{C}^T_{(\underline{x})}$: tangent stiffness matrix.

The damping matrix is an artificial device, which may be introduced such, that the iteration is stable and efficient. The choice of the artificial damping matrix, given in equation (5), leads to an unconditionally stable iteration for geometrically nonlinear problems.

Having solved the problem of iterating in a stable manner from one path to another the problem remains, how to find pathes in the neighbourhood of the existing path, which are relevant for the structure to follow. In some regions of the solution pathes the solutions are sensitive against perturbations. Then small perturbations may lead to a drastic change of the structural response. In such points not necessarily a singularity exists. If in the sensitive region the perturbation of a certain size and direction is applied, the structure snaps to another state of equilibrium.

Size and direction of a critical perturbation is in general not known. However accompanying calculations may give a good estimate for size and direction. With the eigenvalue calculation (6) a set of possible perturbation patterns Φ is calculated.

$$(\underline{C}^T_{x_o} - \lambda_{x_o} \underline{C}^N_{x_o})\ \underline{\Phi}_1 = 0. \tag{6}$$

The distance to a singularity in the direction of the buckling mode Φ_1 of the actual state the can be obtained from (7).

$$(\underline{C}^T_{x_o} - \lambda_\Phi \underline{C}^N_{\Phi_1})\ \underline{\Phi}_2 = 0. \tag{7}$$

λ_Φ gives a measure for the distance to the solution in direction of $\underline{\Phi}_1$. If a perturbation energy is introduced (8), λ_Φ equals the scalar quantity π_p, with the scaling of (9):

$$\pi_p = \underline{p}^T_p\ \Delta \underline{x} \tag{8}$$

$\underline{p}_p$: perturbation vector $= \Phi_1\ C^N_{x_o}$

$\Delta\underline{\bar{x}}$: $\lambda_\Phi \Phi_1 + \Delta\underline{\bar{x}}$; $\underline{p}^T_p\ \Delta\underline{\bar{x}} = 0$

$$\Phi^T_1\ C^N_{x_o} \Phi_1 = \Phi^T_2\ C^N_{\Phi_1}\ \Phi_2 = 1.0. \tag{9}$$

Now the potential energy for the increment may be extended by a constraint, which imposes the perturbation of size to the structure (10):

$$\delta\left(\tfrac{1}{2}\ \Delta\underline{x}_i\ \underline{C}^s_{(\underline{x}_o + \Delta\underline{x}_o)}\ \Delta\underline{x}_i + \lambda(\underline{p}^T_p \Delta\underline{x}_i - \lambda_\Phi)\right) = 0. \tag{10}$$

The iterations with the constraint shift the solution to the neighbourhood of the secondary path. Then the contraint is dropped and the iteration settles at the new solution path. A stable iteration scheme is necessary for this, which may be the above mentioned artificial damping. Examples for the method can be found in /3/.

The above mentioned strategy is related to dynamic perturbation strategy, It may be useful for investigation of secondary pathes and avoids pinpointing on bifurcation points.

References

/1/ Esslinger, M. and Geier, B. "Hochgeschwindigkeitsaufnahmen vom Beulvorgang dünnwandiger Zylinder", Der Stahlbau 39, 73-76, 1970.

/2/ Ramm, E. "Strategies for tracing the nonlinear response near limit points" in Nonlinear Finite Element Analysis in Structural Mechanics, (eds. W. Wunderlich, E. Stein and K.-J. Bathe), Springer Berlin, 63-89, 1980.

/3/ Kröplin, B., Dinkler, D. and Hillmann, J. "An Energy Perturbation Applied To Nonlinear Structural Analysis", Computer Methods in Applied Mechanics And Engineering 52, 885-897, 1985.

/4/ Kröplin, B. "Implicit Relaxation Applied to Postbuckling Analysis of Cylindrical Shells" in Flexible Shells Theory and Applications, (eds. E.L. Axelrad and F.A. Emmerling), Springer Berlin, 163-174, 1984.

DECOMPOSITION METHODS FOR ADHERENCE PROBLEMS IN FINITE ELASTICITY

P. LE TALLEC   A. LOTFI

Service de Mathématiques
LABORATOIRE CENTRAL DES PONTS ET CHAUSSEES
58 Boulevard Lefebvre, 75732 PARIS Cedex 15
France

SUMMARY

The problem discussed in this paper consists in computing the large deformations of incompressible elastic bodies which are glued (not fixed) on part of their boundaries. The proposed numerical technique is based on the augmented Lagrangian approach already used in GLOWINSKI-LE TALLEC [1982] for the numerical solution of two-dimensional equilibrium problems in Finite Elasticity, and is organized as follows :

i) the adhesion problem is first discretized in time, which reduces it to a sequence of contact problems in Finite Elasticity with friction forces ;

ii) each contact problem is then transformed into a saddle-point problem obtained by considering the displacements, the strains and the relative displacement at the contact surface as independent variables ;

iii) these saddle-point problems are finally solved by an iterative technique which considers one variable at a time, thus reducing the global algorithm to the successive solution of a linear elasticity problem with a fixed stiffness matrix, of homogeneous local finite elasticity problems, and of local adhesion problems set on the contact surface.

INTRODUCTION

The purpose of this paper is to describe numerical methods (namely augmented Lagrangian techniques) for solving nonlinear constrained problems involving several vector fields. The strategy is based on operator's splitting and permits an accurate treatment of the constraints. Compared to early applications of these methods (FORTIN-GLOWINSKI [1982][1984]) we introduce a more elaborate splitting strategy and deal with a more general and more constrained situation.

Herein, the method will be illustrated in *three-dimensional finite elasticity* with adhesion and the constraints to be satisfied will be *incompressibility, frame-indifference, unilateral contact*. We first recall the equations of incompressible finite elasticity with adhesion forces (§1), discretize them with respect to time (§2), and then introduce an augmented Lagrangian decomposition of this discretized problem (§3). A numerical algorithm for solving this Lagrangian problem is proposed in §4, and solution procedures are presented respectively for the local incompressible problem in strains (§5) and for the adhesion problem on the contact surface (§6). Numerical results are given in conclusion.

1. EQUATIONS OF THE MODEL PROBLEM

The problem consists in determining at each time t the position $\underset{\sim}{x} + \underset{\sim}{u}(\underset{\sim}{x},t)$ of any particle $\underset{\sim}{x}$ of a given elastic body. This body occupies a known domain Ω in its stress-free reference configuration, is subjected to a given distribution of external loads, is fixed on a part Γ_1 of its boundary Γ, and is glued to a rigid support on another part Γ_c of its boundary (Fig. 1). For handling the large deformations which are involved, we label any particle $\underset{\sim}{x}$ by its position in the reference configuration (Lagrange coordinates) and we relate both $\underset{\sim}{x}$ and its displacement $\underset{\sim}{u}(\underset{\sim}{x},t)$ to a fixed rectangular Cartesian coordinate system. In addition, on the glued surface, we introduce the adhesion intensity $\beta(\underset{\sim}{x},t)$ which measures the proportion of glue links which are still active. Finally, we denote by $\underset{\sim}{f}$ the density of external body forces measured per unit volume of Ω, we denote by $\underset{\sim}{g}$ the density of external surface tractions applied on $\Gamma_2 = \Gamma - \Gamma_1 - \Gamma_c$, and we denote by $\beta_o(\underset{\sim}{x})$ the adhesion intensity at time $t=0$.

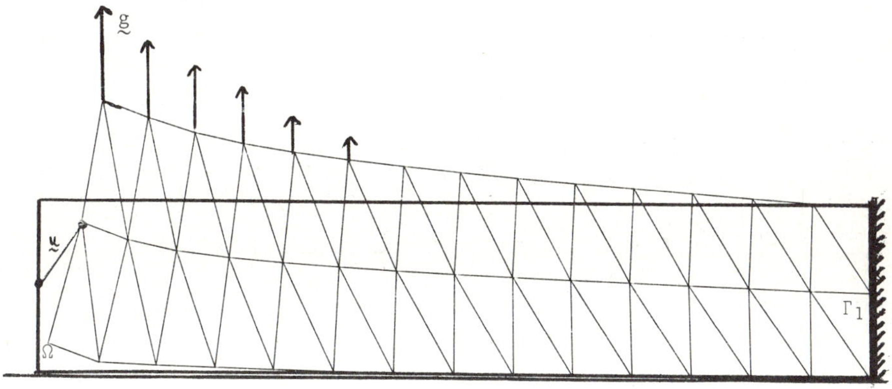

$\Gamma_c : \beta_o = 1$

- Figure 1 -

Reference and deformed configuration.

Neglecting inertia effects, assuming the constitutive material to be incompressible and hyperelastic, supposing the loading $\underset{\sim}{f}$ *and* $\underset{\sim}{g}$ *independant of the displacement* (dead loading), the evolution of $\underset{\sim}{u}$ and β is governed by the following energetic principle :

AT ANY TIME t, IN ANY ADMISSIBLE VIRTUAL PERTURBATION OF THE REAL CONFIGURATION, THE ENERGY DISSIPATED BY BREAKING ADHESION MUST BE BIGGER THAN THE POTENTIAL ENERGY RESTITUTED BY THE STRUCTURE.

Here, by construction, $\beta \in [0,1]$ (β is a proportion of active links), $\underset{\sim}{u} = 0$ on Γ_1 (the body is fixed on Γ_1), $\underset{\sim}{u}\cdot\underset{\sim}{n} \leq 0$ on Γ_c (the body cannot penetrate its rigid support), and $\det(\underline{1}+\nabla\underset{\sim}{u}) = 1$ in Ω (the body is incompressible). Therefore, the set of admissible $\{\underset{\sim}{u},\beta\}$ is finally defined by

$$K = \Big\{\{\underset{\sim}{v},\gamma\} \in \underset{\sim}{H}^1(\Omega) \times L^4(\Gamma_c),\ 0 \leq \gamma \leq 1 \text{ on } \Gamma_c,\ \underset{\sim}{v}\cdot\underset{\sim}{n} \leq 0 \text{ on } \Gamma_c,$$
$$\underset{\sim}{v} = 0 \text{ on } \Gamma_1,\ \det(\underline{1}+\nabla\underset{\sim}{v}) = 1 \text{ a.e. in } \Omega\Big\}. \qquad (1.1)$$

Moreover, the potential energy associated to a displacement field $\underset{\sim}{v}$ and to a distribution of adhesion intensity γ is classically defined by

$$J_{POT}(\underset{\sim}{v},\gamma) = \int_\Omega \mathcal{W}(\underline{1}+\nabla\underset{\sim}{v})\,d\Omega - \int_\Omega \underset{\sim}{f}\cdot\underset{\sim}{v}\,d\Omega - \int_{\Gamma_2} \underset{\sim}{g}\cdot\underset{\sim}{v}\,d\Gamma$$
$$- \int_{\Gamma_c} w\gamma\,d\Gamma + \frac{1}{2\varepsilon}\int_{\Gamma_c} \gamma^2|\underset{\sim}{v}|^2 d\Gamma, \qquad (1.2)$$

with $\mathcal{W}(\cdot)$ the stored elastic energy density, w the Dupré superficial energy and ε a small regularizing factor introduced by FREMOND [1985] in his modeling of adhesion phenomena. Also following FREMOND, we introduce as potential of dissipation the functional

$$J_{DIS}\left(\frac{\partial\beta}{\partial t}\right) = \frac{1}{2}\int_{\Gamma_c}\int_{\Gamma_c} d(\underset{\sim}{x}-\underset{\sim}{y})\,\frac{\partial\beta}{\partial t}(\underset{\sim}{x},t)\,\frac{\partial\beta}{\partial t}(\underset{\sim}{y},t)\,d\underset{\sim}{x}\,d\underset{\sim}{y} \qquad (1.3)$$

whose gradient at value $\frac{\partial\beta}{\partial t}$ will define the energy dissipated by breaking adhesion. From (1.1), (1.2), (1.3), the energetic principle of the previous page takes the mathematical form

> Find $\{\underset{\sim}{u},\beta\} : [0,T] \to K$ such that :
>
> (i) $\beta(\underset{\sim}{x},0) = \beta_o(\underset{\sim}{x})$ on Γ_c ;
>
> (ii) for almost any t in $[0,T]$, there exists a neighborhood $\mathcal{V}(\underset{\sim}{u},\beta)$ of $\{\underset{\sim}{u},\beta\}$ in K such that, for any $\{\underset{\sim}{v},\gamma\}$ in $\mathcal{V}(\underset{\sim}{u},\beta)$ we have (1.4)
>
> $$\nabla J_{DIS}\left(\frac{\partial\beta}{\partial t}\right)\cdot(\gamma-\beta) \geq J_{POT}(\underset{\sim}{u},\beta) - J_{POT}(\underset{\sim}{v},\gamma).$$

Remark : Typical stored energy functions $\mathcal{W}$ and coupling functions d are of the form

$$\mathcal{W}(\underset{\sim}{F}) = C_1(|\underset{\sim}{F}|^2-3) + C_2(|\text{adj } \underset{\sim}{F}|^2 - 3), \qquad (1.5)$$

$$d(\underset{\sim}{y}) = d_o \exp(-d_o|\underset{\sim}{y}|). \qquad (1.6)$$

In (1.5) $\mathcal{W}$ corresponds to the so-called Mooney-Rivlin material and adj $\underset{\sim}{F}$ denotes the transpose of the cofactor matrix of $\underset{\sim}{F}$.

2. IMPLICIT TIME DISCRETIZATION

Let us introduce particular instants $t_o = 0 < t_1 < \dots < t_P = T$ and solve (1.4) only at those instants, replacing the time derivatives $\frac{\partial\beta}{\partial t}(\underset{\sim}{x},t_p)$ by the finite differences

$$\beta(\underset{\sim}{x},t_p) - \beta(\underset{\sim}{x},t_{p-1}) / (t_p - t_{p-1}).$$

In other words, let us proceed to an implicit (backward) time discretization of the time dependent problem (1.4). From the convexity of J_{DIS}, after discretization, and denoting by $\{\underset{\sim}{u}_p,\beta_p\}$ the values of the discretized solution at time t_p, (1.4) becomes

$\forall 1 \leqslant p \leqslant P$, <u>with</u> β_{p-1} <u>known, find a local minimum</u> $\{\underset{\sim}{u}_p,\beta_p\}$ of J_p <u>over the set</u> K <u>of admissible solutions</u>.	(2.1)

In (2.1), K is the set introduced in (1.1), and the functional J_p is defined by

$$J_p(\underset{\sim}{v},\gamma) = (t_p - t_{p-1})\, J_{DIS}\left[\frac{\gamma - \beta_{p-1}}{t_p - t_{p-1}}\right] + J_{POT}(\underset{\sim}{v},\gamma) \qquad (2.2)$$

that is, from (1.2) and (1.3),

$$J_p(\underset{\sim}{v},\gamma) = \underbrace{\int_\Omega \mathcal{W}(\underset{\sim}{1}+\underset{\sim}{\nabla v})\,d\Omega}_{\text{elastic energy}} - \underbrace{\int_\Omega \underset{\sim}{f}\cdot\underset{\sim}{v}\,d\Omega - \int_{\Gamma_2} \underset{\sim}{g}\cdot\underset{\sim}{v}\,d\Gamma}_{\text{dead loading}} + \underbrace{\frac{1}{2\varepsilon}\int_{\Gamma_c} \gamma^2|\underset{\sim}{v}|^2 d\Gamma}_{\text{friction}}$$

$$\underbrace{- \int_{\Gamma_c} w\gamma + \frac{1}{2(t_p - t_{p-1})}\int_{\Gamma_c}\int_{\Gamma_c} d(\underset{\sim}{x}-\underset{\sim}{y})[\gamma-\beta_{p-1}](\underset{\sim}{x})[\gamma-\beta_{p-1}](\underset{\sim}{y})\,d\underset{\sim}{x}\,d\underset{\sim}{y}}_{\text{adhesion}}.$$

As written, the discretized problem (2.1) consists in *minimizing the nonconvex functionals* J_p *over the nonconvex set* K. In view of its numerical solution, (2.1) is first written in an abstract form by introducing

$$V = \{\underset{\sim}{v} \in \underset{\sim}{H}^1(\Omega),\ \underset{\sim}{v} = 0 \text{ on } \Gamma_1\}, \qquad (2.3)$$

$$\mathcal{H} = \{\underset{\sim}{H} = \{\underset{\sim}{F},\underset{\sim}{e},\gamma\} \in L^2(\Omega)^9 \times \underset{\sim}{L}^2(\Gamma_c) \times L^2(\Gamma_c)\} \qquad (2.4)$$

$$\mathcal{G} : \begin{cases} V \to \mathbb{R}, \\ G(\underset{\sim}{v}) = \int_\Omega C_1|\underset{\sim}{1}+\underset{\sim}{\nabla v}|^2 d\Omega - \int_\Omega \underset{\sim}{f}\cdot\underset{\sim}{v}\,d\Omega - \int_{\Gamma_2} \underset{\sim}{g}\cdot\underset{\sim}{v}\,d\Gamma, \end{cases} \qquad (2.5)$$

$$\mathcal{F} : \begin{cases} \mathcal{H} \to \mathbb{R} \cup \{+\infty\}, \\ \mathcal{F}(\underset{\sim}{H}) = +\int_\Omega \{W(\underset{\sim}{F}) - C_1 |\underset{\sim}{F}|^2\} d\Omega - \int_{\Gamma_c} w\gamma \, d\Gamma + \frac{1}{2\varepsilon} \int_{\Gamma_c} \gamma^2 |\underset{\sim}{e}|^2 d\Gamma \\ \quad + \frac{1}{2(t_p - t_{p-1})} \int_{\Gamma_c} \int_{\Gamma_c} d(\underset{\sim}{x} - \underset{\sim}{y}) [\gamma - \beta_{p-1}](\underset{\sim}{x}) [\gamma - \beta_{p-1}](\underset{\sim}{y}) d\underset{\sim}{x} d\underset{\sim}{y} \\ \quad \text{if } \det \underset{\sim}{F} = 1, \ 0 \leqslant \gamma \leqslant 1 \text{ and } \underset{\sim}{e} \cdot \underset{\sim}{n} \leqslant 0, \\ \mathcal{F}(\underset{\sim}{H}) = +\infty \quad \text{if not,} \end{cases} \tag{2.6}$$

$$B : \begin{cases} V \times \mathcal{H} \to L^2(\Omega)^9 \times \underset{\sim}{L}^2(\Gamma_c), \\ B(\underset{\sim}{v}, \underset{\sim}{H}) = \{\underset{\sim}{1} + \nabla \underset{\sim}{v} - \underset{\sim}{F}, \underset{\sim}{v}|_{\Gamma_c} - \underset{\sim}{e}\}. \end{cases} \tag{2.7}$$

With the above notations, (2.1), that is the time-discretized adhesion problem in finite elasticity, takes the usual form

$$\boxed{\begin{array}{l} \forall 1 \leqslant p \leqslant P, \ \underline{\text{with}} \ \beta_{p-1} \ \underline{\text{known}}, \ \underline{\text{find a local minimum of}} \ \mathcal{F}(\underset{\sim}{H}) + \mathcal{G}(\underset{\sim}{v}) \\ \underline{\text{over the space}} \ V \times \mathcal{H} \ \underline{\text{under the linear constraint}} \ B(\underset{\sim}{v}, \underset{\sim}{H}) = 0. \end{array}} \tag{2.8}$$

3. AUGMENTED LAGRANGIAN DECOMPOSITION OF (2.8)

Two strategies can be employed for the numerical solution of (2.8) :

(i) the first one (Penalty + Newton) makes the functional $\mathcal{F}$ differentiable by penalizing the constraints involved in its definition, eliminates the variables $\underset{\sim}{F}$ and $\underset{\sim}{e}$ through the linear constraint $B(\underset{\sim}{v}, \underset{\sim}{H}) = 0$, and obtains afterwards an unconstrained minimization problem with respect to the variable $\{\underset{\sim}{v}, \gamma\}$, problem to be solved by a general purpose Newton's method ;

(ii) the second one, to be presented hereafter, treats the linear constraint $B(\underset{\sim}{v}, \underset{\sim}{H}) = 0$ by an augmented Lagrangian technique (POWELL [1969], HESTENES [1969]). The problem in $\underset{\sim}{v}$ (displacements), the problem in $\underset{\sim}{F}$ (strains) and the problem in $\{\underset{\sim}{e}, \gamma\}$ (adhesion) can then be treated independently.

Classically, augmented Lagrangian methods replace the linearly constraint minimization problem (2.8) by the saddle-point problem

$$\boxed{\begin{array}{l} \forall 1 \leqslant p \leqslant P, \ \underline{\text{with}} \ \beta_{p-1} \ \underline{\text{known}}, \ \underline{\text{solve}} \\ \sup\limits_{\underset{\sim}{\mu} \in L^2(\Omega)^9 \times \underset{\sim}{L}^2(\Gamma_c)} \left\{ \inf\limits_{(\underset{\sim}{v}, \underset{\sim}{H}) \in V \times \mathcal{H}} \mathcal{L}(\underset{\sim}{v}, \underset{\sim}{H}; \underset{\sim}{\mu}) \right\}. \end{array}} \tag{3.1}$$

In (3.1), the supremum and the infimum are local, the spaces V and $\mathcal{H}$ are as defined in (2.3) and (2.4), and the augmented Lagrangian $\mathcal{L}$ is defined by

functional penalty duality

$$\mathcal{L}(\underset{\sim}{v},\underset{\sim}{H};\mu) = \mathcal{F}(\underset{\sim}{H}) + \mathcal{G}(\underset{\sim}{v}) + \frac{R}{2}\|B(\underset{\sim}{v},\underset{\sim}{H})\|^2 - \langle \underset{\sim}{\mu}, B(\underset{\sim}{v},\underset{\sim}{H})\rangle , \tag{3.2}$$

with : $$\|B(\underset{\sim}{v},\underset{\sim}{H})\|^2 = \int_\Omega q(\underset{\sim}{x})\ |\underset{\sim}{1}+\nabla\underset{\sim}{v}-\underset{\sim}{E}|^2 d\Omega + \int_{\Gamma_c} q_c |\underset{\sim}{v}-\underset{\sim}{e}|^2 d\Gamma , \tag{3.3}$$

$$\langle \underset{\sim}{\mu}, B(\underset{\sim}{v},\underset{\sim}{H})\rangle = \int_\Omega \underset{\sim}{\mu}\cdot(\underset{\sim}{1}+\nabla\underset{\sim}{v}-\underset{\sim}{E})d\Omega + \int_{\Gamma_c} \underset{\sim}{\mu}_c\cdot(\underset{\sim}{v}-\underset{\sim}{e})d\Gamma . \tag{3.4}$$

Above, R, $q(\underset{\sim}{x})$ and $q_c(\underset{\sim}{x})$ are arbitrary strictly positive numbers whose choice has no effect on the values of the solutions of (3.1) but determines the speed of convergence of the algorithm used for its numerical solution.

Obviously, if $\{\underset{\sim}{u}_p, \underset{\sim}{G}_p; \underset{\sim}{\lambda}_p\}$ is a solution of (3.1), then $\frac{\partial\mathcal{L}}{\partial\underset{\sim}{\mu}}(\underset{\sim}{u}_p, \underset{\sim}{G}_p; \underset{\sim}{\lambda}_p)$ is equal to zero, thus, from (3.2), $B(\underset{\sim}{u}_p, \underset{\sim}{G}_p)$ is equal to zero and by definition of (3.1), $\{\underset{\sim}{u}_p, \underset{\sim}{G}_p\}$ minimizes locally $(\underset{\sim}{v},\underset{\sim}{H};\underset{\sim}{\lambda}_p)$ over the kernel of B ; in other words $\{\underset{\sim}{u}_p, \underset{\sim}{G}_p\}$ is a solution of (2.8). Moreover, formally, the converse is also true. Therefore, from now on, *instead of solving the adhesion problem in Finite Elasticity under its variational formulation* (2.8) [or (2.1)], *we solve it under its saddle-point* (augmented Lagragian) *formulation* (3.1).

4. UZAWA ALGORITHM

It consists in solving (3.1) with respect to the (dual or stress) variable $\underset{\sim}{\mu}$ by a *steepest descent method*. Taking the step parameter ρ equal to R, and computing the quantity $\underset{\{v,H\}}{\mathrm{INF}}\ \mathcal{L}(\underset{\sim}{v},\underset{\sim}{H};\underset{\sim}{\mu})$ by block-relaxation, the application of this algorithm to (3.1) leads to the following :

β_o := adhesion intensity at time $t = 0$;

choose λ_o in $L^2(\Omega)^9 \times L^2(\Gamma_c)$;

choose $\{F_o, h_o\}$ in $L^2(\Omega)^9 \times L^2(\Gamma_c)$;

$G_o := \{F_o, h_o, \beta_o\}$;

for time step p equal 1 to P, do

$\{G^o, \lambda^o\} = \{G_{p-1}, \lambda_{p-1}\}$;

for iteration n equal 1 until $\left(\|G^n - G^{n-1}\| + \|B(u^n, G^n)\|\right)$ small, do

block relaxation

$$\begin{cases} \text{solve } \mathcal{L}(u^n, G^{n-1}; \lambda^{n-1}) \leq \mathcal{L}(v, G^{n-1}; \lambda^{n-1}), \\ \forall v \in V, \ u^n \in V ; \end{cases} \tag{4.1}$$

$$\begin{cases} \text{solve } \mathcal{L}(u^n, G^n; \lambda^{n-1}) \leq \mathcal{L}(u^n, H, \lambda^{n-1}), \\ \forall H \in \mathcal{H}, \ G^n \in \mathcal{H} ; \end{cases} \tag{4.2}$$

gradient update

$$\lambda^n := \lambda^{n-1} - RqB(u^n, G^n) ; \tag{4.3}$$

end loop n

$$\{u_p, G_p; \lambda_p\} = \{u^n, G^n; \lambda^n\} ; \tag{4.4}$$

end loop p.

Obviously, many variants can be imagined for such algorithms, some of them being described in FORTIN-GLOWINSKI [1982, 1984 : chap. 3]. Moreover, they can be interpreted as alternating directions time-integrators of a parabolic evolution equation in λ, in the sense introduced by PEACEMAN-RACHFORD [1955] and studied by LIONS-MERCIER [1979]. But, what is more important is to understand to which numerical operators correspond the different steps of algorithm (4.1)-(4.4).

First, from the definitions of $\mathcal{G}$, $\mathcal{L}$ and B, (4.1) is

$$a(u^n, v) = L(v), \quad \forall v \in V, \ u^n \in V, \tag{4.5}$$

with

$$a(u^n, v) = \int_\Omega C_1 \nabla u^n \cdot \nabla v \, d\Omega + R\int_\Omega q \nabla u^n \cdot \nabla v \, d\Omega + R\int_{\Gamma_c} q_c u^n \cdot v \, d\Gamma,$$

$$L(v) = \int_\Omega f \cdot v \, d\Omega + \int_{\Gamma_2} g \cdot v \, d\Gamma - \int_\Omega C_1 \operatorname{div} v \, d\Omega$$
$$+ \int_\Omega \{Rq(F^{n-1} - 1) + \lambda^{n-1}\} \cdot \nabla v \, d\Omega + \int_{\Gamma_c} \{Rq_c \, e^{n-1} + \lambda_c^{n-1}\} \cdot v \, d\Gamma .$$

This is the variational formulation of a linear, strongly elliptic, non-homogeneous Poisson equation which, after finite element discretization of V, reduces to a well-posed linear system in $\mathbb{R}^{\dim V_h}$. The associated matrix is sparse, positive definite, uncoupled in each direction, does not change during the iterative process, and can thus be constructed and factorized once for all. The linear system can then be solved very efficiently at each step in $O(\dim V_h)$ operations.

Next, from the construction of $\mathcal{F}$, $\mathcal{L}$ and B, (4.2) splits into

$$\int_\Omega J_E(\underset{\sim}{F}^n)d\Omega \leq \int_\Omega J_E(\underset{\sim}{F})d\Omega, \quad \forall \underset{\sim}{F} \in K_E, \quad \underset{\sim}{F}^n \in K_E, \tag{4.6}$$

$$\int_{\Gamma_c} J_c(\underset{\sim}{h}^n,\beta^n)d\Gamma \leq \int_{\Gamma_c} J_c(\underset{\sim}{e},\gamma)d\Gamma, \quad \forall\{\underset{\sim}{e},\gamma\} \in K_c, \quad \{\underset{\sim}{h}^n,\beta^n\} \in K_c . \tag{4.7}$$

The constrained minimization problems (4.6) and (4.7) are local and will be studied respectively in §5 and §6, under the notations

$$J_E(\underset{\sim}{F}) = \{W(\underset{\sim}{F}) - C_1|\underset{\sim}{F}|^2 + \frac{R}{2}q|\underset{\sim}{F}-\underset{\sim}{1}-\nabla\underset{\sim}{v}^n|^2 + \underset{\sim}{\lambda}^{n-1}\cdot\underset{\sim}{F}\} \tag{4.8}$$

$$K_E = \{\underset{\sim}{F} \in (L^2(\Omega))^9, \ \det \underset{\sim}{F} = 1\}, \tag{4.9}$$

$$J_c(\underset{\sim}{e},\gamma) = \{\frac{R}{2}q_c|\underset{\sim}{u}^n-\underset{\sim}{e}|^2 + \underset{\sim}{\lambda}_c^{n-1}\cdot\underset{\sim}{e} + \frac{\gamma^2}{2\varepsilon}|\underset{\sim}{e}|^2 - w\gamma$$
$$+ \frac{1}{2(t_p - t_{p-1})}\int_{\Gamma_c} d(\underset{\sim}{x}-\underset{\sim}{y})[\gamma-\beta_{p-1}](\underset{\sim}{x})[\gamma-\beta_{p-1}](\underset{\sim}{y})d\underset{\sim}{y}\} \tag{4.10}$$

$$K_c = \{\{\underset{\sim}{e},\gamma\}, \ 0 \leq \gamma \leq 1, \ \underset{\sim}{e}\cdot\underset{\sim}{n} \leq 0\}. \tag{4.11}$$

Finally, (4.3) is an explicit gradient update of $\underset{\sim}{\lambda}$ and (4.4) simply expresses that if $\underset{\sim}{G}^n = \underset{\sim}{G}^{n-1}$ is a local minimizer of $\mathcal{L}(\underset{\sim}{u}^n,\cdot;\underset{\sim}{\lambda}^{n-1})$, if $\underset{\sim}{u}^n$ is a local minimizer of $\mathcal{L}(\cdot,\underset{\sim}{G}^{n-1};\underset{\sim}{\lambda}^{n-1})$ and if $B(\underset{\sim}{u}^n,\underset{\sim}{G}^n) = 0$, then $\{\underset{\sim}{u}^n,\underset{\sim}{G}^n\}$ will be the desired solution of the finite elasticity problem (2.8) at time-step p.

From these remarks, algorithm (4.1)-(4.4) reduces to the flow chart represented in Fig. 2.

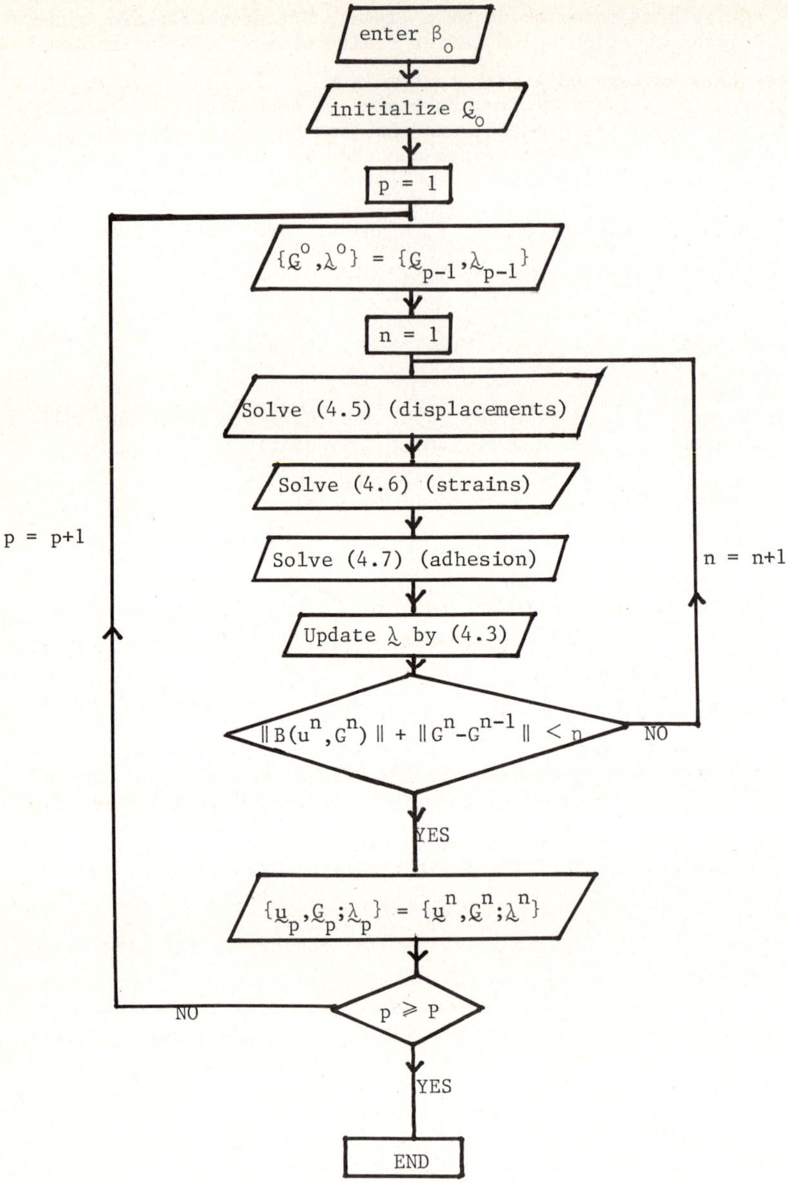

Fig. 2. Flow Chart of Uzawa algorithm (4.1)-(4.4)

5. SOLVING PROBLEM (4.6) IN STRAINS (isotropic case)

By construction, the spatial derivatives of the unknown strains $\underset{\sim}{F}$ do not appear in our problem. Therefore, it is legitimate to use piecewise constant finite elements to approximate $\underset{\sim}{F}$, that is to replace in (4.6) the set K_E by

$$K_E^h = \{\underset{\sim}{E},\ \det \underset{\sim}{E} = 1,\ \underset{\sim}{E}|_{\Omega^e} = \text{constant},\ \forall e = 1, N_h\}, \tag{5.1}$$

where the family $(\Omega^e)_{e=1,N_h}$ forms a regular triangulation of the domain Ω.

In this case, since the minimum value of the sum of independent terms

$$\int_\Omega J_E(\underset{\sim}{E})\,d\Omega = \sum_{e=1}^{N_h} \int_{\Omega^e} J_E(\underset{\sim}{E})\,d\Omega = \sum_{e=1}^{N_h} J_E(\underset{\sim}{E})\ \text{meas}(\Omega^e)$$

is equal to the sum of the minimum value of each term, (4.6) reduces to

$$\forall e = 1,N_h,\ J_E(\underset{\sim}{F}) \leqslant J_E(\underset{\sim}{E}),\ \forall \underset{\sim}{E} \in \mathbb{R}^9 \text{ with } \det \underset{\sim}{E} = 1,\ \det \underset{\sim}{F} = 1, \tag{5.2}$$

with

$$J_E(\underset{\sim}{F}) = \left(\frac{R}{2}\, q - C_1\right) |\underset{\sim}{E}|^2 + \mathcal{W}(\underset{\sim}{E}) - [Rq\,(\underset{\sim}{1} + \underset{\sim}{\nabla}\underset{\sim}{u}^n) - \underset{\sim}{\lambda}^{n-1}] \cdot E \tag{5.3}$$

q, C_1, $\underset{\sim}{u}^n$ and $\underset{\sim}{\lambda}^n$ being averaged over Ω^e. Moreover, if the constitutive hyperelastic material is isotropic, the stored elastic energy density $\mathcal{W}$ only depends on the singular values of $\underset{\sim}{E}$ (frame indifference + symmetry).

From this frame indifference and symmetry, and from the incompressibility, (5.2) can in turn be solved through a change of variables which reduces it to *unconstrained minimization problems in two scalar variables*.

Indeed, let us diagonalize $Rq(\underset{\sim}{1} + \underset{\sim}{\nabla}\underset{\sim}{u}^n) - \underset{\sim}{\lambda}^{n-1}$ into the product $\underset{\sim}{Q}\ \underset{\sim}{D}\ \underset{\sim}{R}$, $\underset{\sim}{Q}$ and $\underset{\sim}{R}$ being orthogonal 3×3 matrices with positive determinant and $\underset{\sim}{D}$ being diagonal and let us set

$$\underset{\sim}{F} = \underset{\sim}{Q}\,\underset{\sim}{T}\,\underset{\sim}{R}\ . \tag{5.4}$$

With respect to the new variable $\underset{\sim}{T}$, (5.2) becomes

$$\forall e = 1,N_h,\quad J_T(\underset{\sim}{T}) \leqslant J_T(\underset{\sim}{E}),\ \forall \underset{\sim}{E} \in \mathbb{R}^9 \text{ with } \det \underset{\sim}{E} = 1,\ \det \underset{\sim}{T} = 1 \tag{5.5}$$

$$J_T(\underset{\sim}{E}) = \left(\frac{R}{2}\, q - C_1\right) |\underset{\sim}{E}|^2 + \mathcal{W}(\underset{\sim}{E}) - \sum_{i=1}^{3} D_{ii}\, E_{ii}\,. \tag{5.6}$$

It is a simple exercise to observe that, in the isotropic case, (5.5) is well-posed and has diagonal matrices $\underset{\sim}{T}$ as solutions. In other words, we can restrict (5.5) to the set of diagonal matrices with determinant equal to 1, set parametrized by $\mathbb{R}^2$ through the mapping

$$\underset{\sim}{T}(t_1, t_2) = \begin{pmatrix} t_1 & 0 & 0 \\ 0 & t_2 & 0 \\ 0 & 0 & 1/t_1 t_2 \end{pmatrix}. \tag{5.7}$$

Then, finally, (5.2) is equivalent to

$$\forall e = 1,N_h,\ J_T(\underset{\sim}{T}(t_1,t_2)) \leqslant J_T(\underset{\sim}{T}(e_1,e_2)),\ \forall\{e_1,e_2\} \in \mathbb{R}^2, \tag{5.8}$$

with J_T and $\underset{\sim}{T}$ respectively given by (5.6) and (5.7).

In summary, based on (5.8), *the numerical procedure for solving problem* (4.6) *in strains for the isotropic case proceeds element by element, and on each element* Ω^e *does the following :*

.*diagonalize* $Rq(\underset{\sim}{1}+\underset{\sim}{\nabla}\underset{\sim}{u}^n) - \underset{\sim}{\lambda}^{n-1}$ *into* $\underset{\sim}{Q}\underset{\sim}{D}\underset{\sim}{R}$;

.*solve locally the unconstrained two-dimensional minimization problem* (5.8) by a Newton's method (very simple to code and efficient since there are only two unknown t_1 and t_2) ;

.*set* $\underset{\sim}{F} = \underset{\sim}{Q}\underset{\sim}{T}(t_1,t_2)\underset{\sim}{R}$ (which automatically respects symmetry, frame-indifference and incompressibility, since $\underset{\sim}{F}$ has the same singular vectors as the local loading $Rq(\underset{\sim}{1}+\underset{\sim}{\nabla}\underset{\sim}{u}^n)-\underset{\sim}{\lambda}^{n-1}$ and since $\underset{\sim}{T}$ was constructed with $\det \underset{\sim}{T} = 1$).

6. SOLVING PROBLEM (4.7) FOR ADHESION

Let us approximate the adhesion intensity β and the displacement $\underset{\sim}{h}$ of the contact surface by piecewise constant finite elements Γ^e defined on Γ_c. If we proceed by relaxation, that is we iteratively suppose that the solution $\{\underset{\sim}{h},\beta\}$ is known on each finite element except on the finite element Γ^i, Problem (4.7) for adhesion reduces to a sequence of minimization problems of the form

$$J_c(\underset{\sim}{h},\beta) \leqslant J_c(\underset{\sim}{e},\gamma),\ \forall\{\underset{\sim}{e},\gamma\} \in K_c \cap \mathbb{R}^4,\ \{\underset{\sim}{h},\beta\} \in K_c \cap \mathbb{R}^4. \tag{6.1}$$

In (6.1), K_c has been defined in (4.11) and we set

$$J_c(\underset{\sim}{e},\gamma) = \frac{R}{2} q_c |\underset{\sim}{e}|^2 - \underset{\sim}{a}\cdot\underset{\sim}{e} + \frac{1}{2\varepsilon}\gamma^2|\underset{\sim}{e}|^2 + \frac{aa}{2}\gamma^2 - b\gamma, \tag{6.2}$$

$$\underset{\sim}{a} = (Rq_c\, \underset{\sim}{u}^n - \underset{\sim}{\lambda}_c^{n-1} ; \tag{6.3}$$

$$aa = d(0)\ \mathrm{meas}(\Gamma^i)/(t_p - t_{p-1}), \tag{6.4}$$

$$\begin{aligned} b = w - \sum_{e\neq i} d(\underset{\sim}{x}_i-\underset{\sim}{x}_e)\ [\beta_p(\underset{\sim}{x}_e) - \beta_{p-1}(\underset{\sim}{x}_e)\ \mathrm{meas}(\Gamma^e)/(t_p - t_{p-1}) \\ + d(0)\ \beta_{p-1}(\underset{\sim}{x}_i)\mathrm{meas}(\Gamma^i)/(t_p - t_{p-1}). \end{aligned} \tag{6.5}$$

As in §5, we can simplify (6.1) and reduce it to a one-dimensional local minimization problem over the closed interval [0,1]. Indeed, if we write the optimality conditions associated to (6.1), we observe that, at the solution, $\underset{\sim}{h}$ is given from β through the relation

$$\underset{\sim}{h} = \varepsilon\underset{\sim}{a}^-/(\beta^2+Rq_c\varepsilon), \tag{6.6}$$

$$\begin{aligned} \underset{\sim}{a}^- &= \underset{\sim}{a} \text{ if } \underset{\sim}{a}\cdot\underset{\sim}{n} \leqslant 0, \\ &= \underset{\sim}{a} - \underset{\sim}{a}\cdot\underset{\sim}{n}\ \underset{\sim}{n} \text{ if not.} \end{aligned} \tag{6.7}$$

Plugging this value of $\underset{\sim}{h}$ back in (6.1), the local adhesion problem (6.1) becomes

Find a local minimum of $\{aa\gamma^2 - 2b\gamma - \varepsilon\|\underset{\sim}{a}^-\|^2/(\gamma^2 + Rq_c\varepsilon)^2\}$ *over the closed interval* $[0,1]$.	(6.8)

In summary, based on (6.8), *the numerical procedure for solving Problem* (4.7) *for adhesion proceeds by relaxation on the finite elements* Γ^e *and solves each local problem* (6.1) *by*

* *computing* $\underset{\sim}{a}^-$ by (6.7),
* *minimizing* (6.8) over $[0,1]$ by a scalar Newton's method, which gives $\beta^n(\underset{\sim}{x}_i)$,
* *computing* $\underset{\sim}{h}$ by (6.6).

7. NUMERICAL RESULTS

We first begin by testing the numerical method presented in §4 to §6 on a problem solved in small strains by GUIDOUCHE and POINT [1985]. The considered body is a hyperelastic compressible bar, occupying the domain $[0,1]\times[0,0.20]\times\mathbb{R}$ in its reference configuration, initially glued on the surface $x_2 = 0$ $(\beta_o = 1)$, and subjected to a vertical surface traction of density $g_2 = 10(1-x_1)$kpa (Fig. 1). The other physical data are

$$w = 0.01 \quad \text{SI},$$

$$d(\underset{\sim}{y}) = d_o \exp(-d_o|\underset{\sim}{y}|), \quad d_o = 1\,000.0 \ \text{SI},$$

$$\varepsilon = 10^{-9},$$

$$\mathcal{W}(\underset{\sim}{F}) = C_1(|\underset{\sim}{F}|^2-3) + C_3(\det \underset{\sim}{F})^2 - C_3 - 2(C_1+C_3)\ln(\det \underset{\sim}{F}),$$

$$C_1 = C_3 = 187.5 \ \text{kpa}.$$

In small strains, the above hyperelastic material corresponds to a classical linearly elastic material with a Young modulus of 1000 kpa and a Poisson coefficient of 0.33. Our computation is performed in plane strains with a variant of algorithm (4.1)-(4.4), taking $R = 8.0$, $q = 2C_1$, $q_c = 1/\varepsilon$, $t_p - t_{p-1} = 5$s, using 201 nodes for constructing the finite element approximation of $\underset{\sim}{u}$, 168 triangles for constructing the piecewise constant approximation of $\underset{\sim}{F}$ and 28 segments for constructing the piecewise constant approximation of β. The variant of algorithm (4.1)-(4.4) consists in using for each n two iterations of block relaxation (4.1)-(4.2) instead of one. The finite elements used for approximating displacements are the asymmetric elements of RUAS [1981] which are well-adapted to incompressible materials. The obtained numerical solution is represented in Fig. 3, is quite similar to the small strains one of GUIDOUCHE and POINT, and corresponds to roughly 1h of CPU time on a VAX 750.

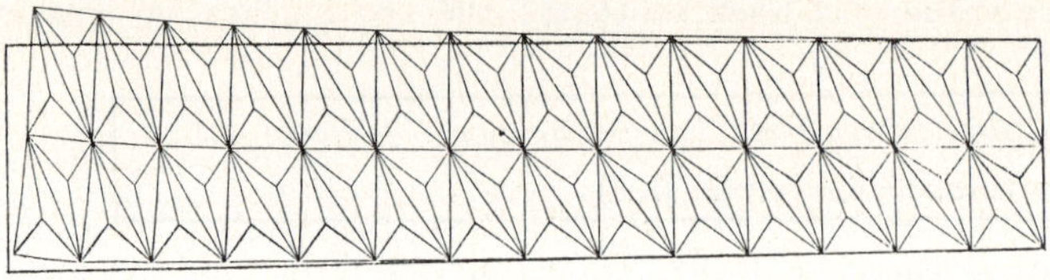

- Figure 3 -

Next, to observe large strains, we consider the same numerical example, but we reduce the thickness of the bar to 0.05m and use a truly incompressible Mooney-Rivlin material with C_1 = 187.5 kpa and C_2 = 0. The adhesion intensity and the vertical displacement of the contact surface are indicated on Fig. 4 for a time t = 800 s.

REFERENCES

FORTIN, M. ; GLOWINSKI, R. (eds) [1982,1984] : Méthodes de Lagrangien Augmenté. Dunod Bordas. Paris. Translated into english at North-Holland, Amsterdam.

FREMOND, M. [1985] : Adhérence des solides. Note to CRAS, 300, Série II, 711-714.

GLOWINSKI, R. ; LE TALLEC, P. [1982] : Numerical solution of problems in incompressible finite elasticity by Augmented Lagrangian methods. SIAM J. Appl. Math., Vol. 42, 2, 400-429.

GUIDOUCHE, H. ; POINT, N. [1985] : Unilateral contact with adherence. Free Boundary Problems, BOSSAVIT and al. eds, Pitman, London.

HESTENES, M. [1969] : Multiplier and gradient methods. J. Optim. Theory and Appl., 4, 303-320.

LIONS, P.L. ; MERCIER, B. [1979] : Splitting algorithms for the sum of 2 nonlinear operators. SIAM J. Numer. Anal., 16, 964-979.

PEACEMAN-RACHFORD [1955] : The numerical solution of parabolic and elliptic differential equations, SIAM J. Appl. Math, 3, 24-41.

POWELL, M.J.D. [1969] : A method for nonlinear constraints in minimization problems. In optimization, FLETCHER ed., chap. 19, Academic Press, N.Y.

RUAS, V. [1981] : "A class of asymmetric simplicial finite elements for solving finite incompressible elasticity problems". Comp. Meth. Appl. Eng., 27, 3, 319-343.

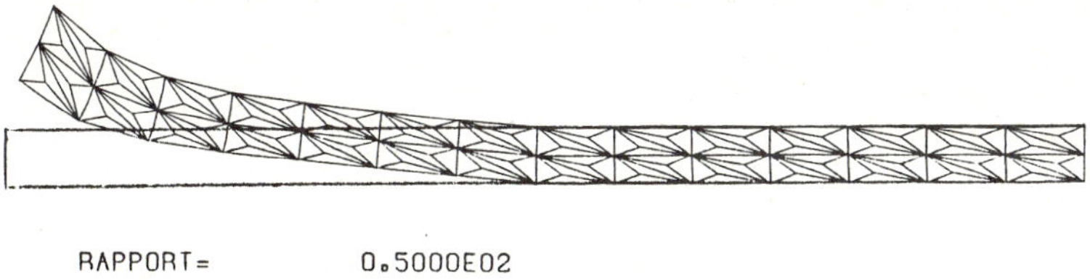

deformed configuration

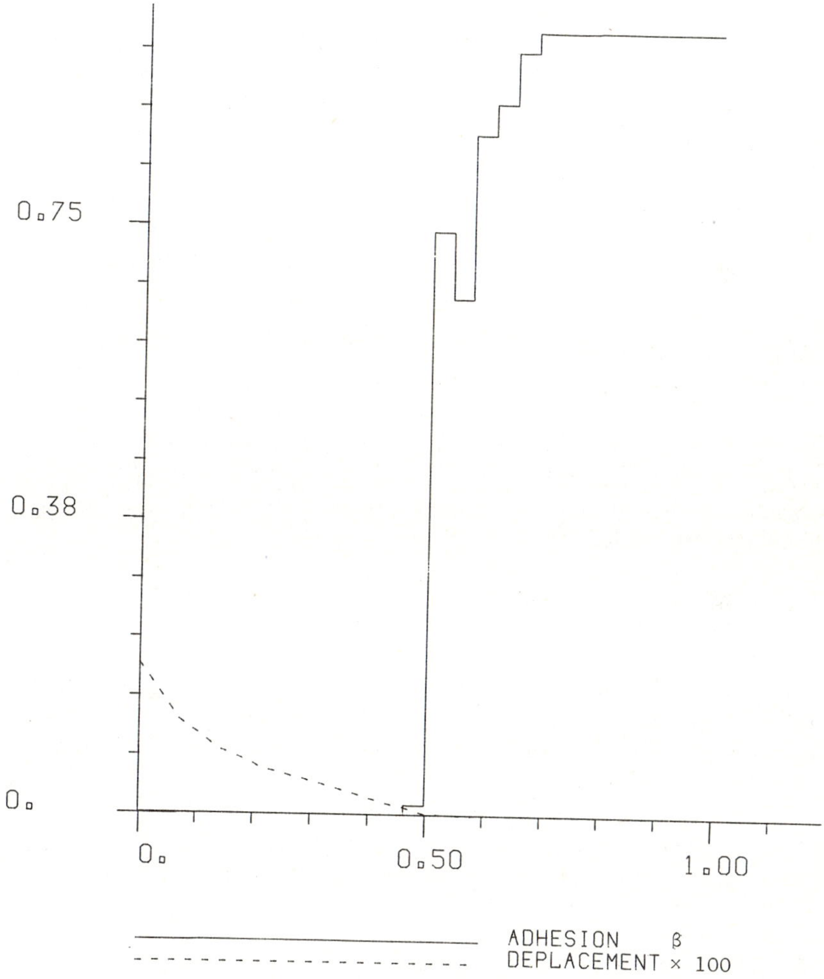

- Figure 4 -

AN ALGORITHM THAT EXPLOITS SYMMETRIES IN BIFURCATION PROBLEMS

H. D. Mittelmann, B. H. Thomson
Department of Mathematics
Arizona State University
Tempe, Arizona 85282, U.S.A.

SUMMARY

Frequently bifurcations in nonlinear eigenvalue problems are due to symmetries in the problem. At bifurcation points the symmetries in the solution are typically reduced on the bifurcating branches. We present an algorithm that by making explicit use of the symmetry behaviour of the solutions allows us to determine these in a reliable and efficient way. Numerical results are presented for a finite-difference discretization of a Duffing equation with periodic boundary conditions.

1. INTRODUCTION

One class of nonlinear parameter-dependent boundary value problems whose discretizations in general inherit bifurcation properties from the continuous case are problems with symmetries. We analyze the behaviour of a generalized inverse iteration method for the numerical solution of these problems near a symmetry-breaking pitchfork bifurcation point. The algorithm is adapted to the symmetry properties present on the different branches. This allows us to guarantee that points on a certain branch are computed. But it also represents a very efficient way to compute these solutions.

Let the nonlinear eigenvalue problem

$$g(u,\lambda) = 0 \tag{1.1}$$

be given, where $g\colon X \times R \to X$, X a Banach space. An algorithm for the augmented system

$$\begin{aligned} g(u,\lambda) &= 0, \\ \|u\|^2 - \rho^2 &= 0 \end{aligned} \tag{1.2}$$

was introduced in [3]. We thus parametrize the solution u in addition to λ by a functional of it as, for example, a norm. In generalization of [3] the algorithm of generalized inverse iteration solves (1.1) starting from a solution $u^{(0)} \in X$, $\|u^{(0)}\| = \rho$ by computing the sequence $(u^{(k)}, \lambda^{(k)})$ according to

$$u^{(k)} = \rho \frac{\tilde{u}^{(k-1)}}{\|\tilde{u}^{(k-1)}\|}, \quad u^{(k)*} g(u^{(k)}, \lambda^{(k)}) = 0,$$

$$\tilde{u}^{(k-1)} = u^{(k-1)} + \delta u^{(k-1)}, \quad k = 1,2,\ldots \text{ where} \tag{1.3}$$

$$\begin{bmatrix} g_u^{(k-1)} & g_\lambda^{(k-1)} \\ u^{(k-1)*} & 0 \end{bmatrix} \begin{bmatrix} \delta u^{(k-1)} \\ * \end{bmatrix} = - \begin{bmatrix} g^{(k-1)} \\ 0 \end{bmatrix}.$$

Here $u^* \in X'$ is such that $u^* u = \|u\|$, $g^{(k-1)} = g(u^{(k-1)}, \lambda^{(k-1)})$ and $g_u^{(k-1)}$, $g_\lambda^{(k-1)}$ are the Frechét derivatives evaluated at the same arguments.

In addition to exhibiting rapid local convergence the above algorithm proved to be robust in the sense that relatively large step sizes along solution branches lead to convergence of (1.3) for a very simple predictor. Proceeding from the (norm -) level ρ to $\rho + \delta\rho$ the solution $u(\rho)$ was simply renormalized. For a generally applicable algorithm, of course, different and more elaborate predictor techniques would be used. It may also be preferable to use Newton's method on the augmented system as corrector iteration. Both these ideas were realized and we refer to [5] and the literature given there.

Here we concentrate on the symmetries present in the solution. Therefore we only consider the basic algorithm (1.3) and suggest a modification of it along the lines of [4] to exploit these symmetries.

After defining symmetry-breaking bifurcation points in the next section we analyze the regularity of the corrector Jacobian in these points. In section 4 a modified generalized inverse iteration is proposed and subsequently analyzed. A nontrivial example is described in some detail in section 5. It is one-dimensional but exhibits an interesting bifurcation behaviour and serves to demonstrate the algorithm better than the classical constructed example from [6] used in [4]. Section 6 contains the numerical implementation of the method for this problem and in the last section some results are presented.

2. SYMMETRY-BREAKING BIFURCATION POINTS

In a simple bifurcation point (u_0, λ_0) of (1.1) we have

$$N(g_u^0) = \text{span}\,\{\phi_0\}, \quad \|\phi_0\| = 1,$$

$$R(g_u^0) \text{ is closed and codim } R(g_u^0) = 1. \tag{2.1}$$

$g_u^0 = g_u(u_0,\lambda_0)$ is thus a Fredholm operator of index zero and its adjoint operator satisfies

$$N(g_u^{0*}) = \text{span}\ \{\phi_0^*\},\ R(g_u^0) = \{x \mid \phi_0^* x=0\}\ . \tag{2.2}$$

In addition we assume the zero eigenvalue of g_u^0 to have algebraic multiplicity one and thus we may take

$$\phi_0^*\phi_0 = 1. \tag{2.3}$$

If (u_0,λ_0) is a pitchfork bifurcation point then

$$\begin{aligned} &\phi_0^* g_\lambda^0 = 0,\ \phi_0^* g_{uu}^0 \phi_0\phi_0 = 0, \\ &\phi_0^*(g_{u\lambda}^0\phi_0 + g_{uu}^0 v_0\phi_0) \neq 0. \end{aligned} \tag{2.4}$$

Where v_0 is the unique solution of

$$g_u^0 v_0 + g_\lambda^0 = 0,\ \phi_0^* v_0 = 0. \tag{2.5}$$

We now assume that there is a finite group $\gamma \in L\ (X)$ such that

$$g(Tu,\lambda) = Tg(u,\lambda),\ \text{for all}\ T \in \gamma,\ u \in X,\ \lambda \in R. \tag{2.6}$$

The subspace

$$X_s\ :=\ \{u\varepsilon X \mid u=Tu,\ \text{for all}\ T\varepsilon\gamma\} \tag{2.7}$$

contains γ-symmetric elements and is invariant under $g(\cdot,\lambda)$. A typical case is $\gamma = \{I, S\}$, where I is the identity and $S \in L\ (X)$ such that $S^2 = I$, $S \neq I$. This case was considered in [4]. A more general case with three nontrivial symmetries will be solved below. Let S denote any such symmetry then the space X can be split as

$$X = X_s \oplus X_a \tag{2.8}$$

where

$$X_s = \{x \varepsilon X \mid Sx=x\}, \qquad X_a = \{x \varepsilon X \mid Sx=-x\} \tag{2.9}$$

and the analogue holds for the dual spaces.

The pitchfork bifurcation point (u_0,λ_0) of (1.1) is called symmetry-breaking, if

$$u_0 \ \varepsilon \ X_s, \ \phi \ \varepsilon \ X_a. \tag{2.10}$$

That is on one of the branches bifurcating from this point the symmetry w.r.t. $S \ \varepsilon \ \gamma$ present in (u_0,λ_0) is lost. This can be seen more explicitly from a local representation of the two branches intersecting at (u_0,λ_0)

$$\begin{aligned} &u_1(\xi) = u_0 + w_1(\xi), && \lambda_1(\xi) = \lambda_0 + \xi, \ \xi \ \varepsilon \ [-\xi_0, \xi_0], \\ &u_2(\xi) = u_0 + \xi\phi_0 + w_2(\xi), && \lambda_2(\xi) = \lambda_0 + O(\xi^2), \\ &\|w_i(\xi)\| = O(\xi^i), \ i = 1, 2, && w_1(\xi) \ \varepsilon \ X_s. \end{aligned} \tag{2.11}$$

In addition we assume

$$\overset{*}{\phi_0} \ \varepsilon \ (X')_a$$

which, however, is always the case for finite-dimensional X.

3. REGULARITY OF THE JACOBIAN

Introducing the notation $F(y)$, $y = (u,\lambda)$, for the augmented system (1.2) and thus

$$F_y = \begin{bmatrix} g_u & g_\lambda \\ u^* & 0 \end{bmatrix} \tag{3.1}$$

for the Jacobian in (1.3) we now address the question of the regularity of F_y along the solution branch.

We note that F_y is regular in regular points, i.e. points in which g_u is regular, and in turning points w.r.t. λ that are not simultaneously turning points in ρ, i.e. $\frac{d\rho}{ds} = 0$ holds where s denotes arclength (cf. [5]). We thus concentrate on simple symmetry-breaking bifurcation points in which F_y is singular but it is the 'degree' of singularity we have to determine.

The following lemma is the analogue of Theorem 5.1 in [2]. We assume that u and λ can be parametrized by ρ in the neighbourhood of the simple bifurcation point $y_0 = (u(\rho_0), \lambda(\rho_0))$.

<u>Lemma 3.1</u> Let F(y) be twice continuously differentiable with respect to y and let $F_y(y_0)$ be a Fredholm operator of index zero with a simple zero eigenvalue and

$$F_y(y_0)\Phi_0 = 0, \quad F_y^*(y_0)\Phi_0^* = 0 .$$

Then for $|\rho-\rho_0| < \delta$, some $\delta < 0$, there exists a twice continuously differentiable pair $(\beta(\rho), \Phi(\rho))$ such that

$$F_y(y(\rho))\Phi(\rho) = \beta(\rho)\Phi(\rho),$$
$$\beta(\rho_0) = 0, \quad \Phi(\rho_0) = \Phi_0. \tag{3.2}$$

<u>Proof</u> We consider the system of equations

$$H(\beta, \Phi, \rho) = \begin{bmatrix} F_y(y(\rho))\Phi - \beta\Phi \\ \Phi_0^*\Phi_0 - 1 \end{bmatrix} = 0.$$

This system has a solution $(0, \Phi_0, \rho_0)$ and its Frechét derivative there with respect to (Φ, β) is

$$H'(0, \Phi_0, \rho_0) = \begin{bmatrix} F_y(y_0) & \Phi_0 \\ \Phi_0^* & 0 \end{bmatrix} .$$

This operator is nonsingular by Lemma I in [2] and thus the result follows from the implicit function theorem.

<u>Theorem 3.2</u> Let $F(u,\lambda)$ be twice continuously differentiable with respect to u and λ, let $(u_0, \lambda_0) = (u(\rho_0), \lambda(\rho_0))$ be a simple bifurcation point and let $u_0^*v_0 \neq 0$, v_0 as in (2.5). Then, for some $\delta > 0$, $K > 0$ and $0 < |\rho - \rho_0| < \delta$

$$\|F_y^{-1}(u(\rho),\lambda(\rho))\| \leq \frac{K}{|\rho-\rho_0|^p} \tag{3.3}$$

where $p = 1$ if $\dot{\lambda}_0 = \frac{d\lambda}{d\rho}(\rho_0) \neq 0$ and $p \geq 2$ otherwise.

Proof Differentiating (1.1) with respect to ρ and evaluating at ρ_0 yields

$$g_u^0\,\dot{u}_0 + g_\lambda^0\dot{\lambda}_0 = 0. \tag{3.4}$$

If $\dot{\lambda}_0 \neq 0$ we have $\dot{u}_0 = \dot{\lambda}_0\,v_0$ from (2.5). Differentiating (3.2) with respect to ρ, evaluating at ρ_0 and multiplying by Φ_0^* we obtain

$$\overset{*}{\Phi}_0F_y^0\,\dot{\Phi}_0(\rho_0) + \overset{*}{\Phi}_0F_{yy}^0\dot{y}(\rho_0)\Phi_0 = \beta(\rho_0)\overset{*}{\Phi}_0\dot{\Phi}(\rho_0) + \dot{\beta}(\rho_0)\overset{*}{\Phi}_0\Phi_0 \tag{3.5}$$

We first determine the eigenvector Φ_0 of Lemma 3.1.

$$F_y^0\Phi_0 = 0,\quad \Phi_0 = \begin{bmatrix} v \\ \alpha \end{bmatrix}$$

is equivalent to

$$g_u^0v + \alpha g_\lambda^0 = 0,\quad \overset{*}{u}_0v = 0.$$

The first equation has the solution $v = \alpha v_0 + \gamma\phi_0$, $\gamma \in R$. We have $\overset{*}{u}_0 \in (X')_s$ and hence $\overset{*}{u}_0\,w = \overset{*}{u}_0\,S^2w = (\overset{*}{u}_0S)(Sw) = -\overset{*}{u}_0w$ for $w \in X_a$. The second equation then yields $0 = \overset{*}{u}_0v = \alpha\overset{*}{u}_0v_0$ and thus $\alpha = 0$. Thus $N(F_y^0) = \text{span}\,\{\Phi_0\}$, $\Phi_0 = \begin{smallmatrix}\phi_0\\0\end{smallmatrix}$ and $R(F_y^0) = \{y \mid \overset{*}{\Phi}_0y=0\}$.
From (2.1)-(2.3), (3.2), (3.4), we have

$$\dot{\beta}(\rho_0) = \overset{*}{\Phi}_0F_{yy}^0\dot{y}(\rho_0)\Phi_0$$

$$= \dot{\lambda}_0\overset{*}{\phi}_0(g_{uu}^0v_0+g_{u\lambda}^0)\,\phi_0.$$

In view of (2.4) the eigenvalue $\beta(\rho)$ in (3.2) decreases as $|\rho - \rho_0|$ for $\rho \to \rho_0$ if $\dot{\lambda}_0 \neq 0$ while this decrease is at least quadratic if $\dot{\lambda}_0 = 0$.

From this (3.3) follows. This behaviour corresponds to that of the Jacobian g_u (cf. [4]) and hence the augmenting by the normalizing condition in (1.2) has not improved the regularity.

4. A METHOD EXPLOITING SYMMETRIES

Bifurcation from the trivial solution of a nonlinear eigenvalue problem may be interpreted as symmetry-breaking bifurcation in the sense of section 2. by taking $\gamma = \{I, -I\}$, $X_s = \{0\}$, $X_a = X$. The basic generalized inverse iteration (1.3) introduces in this case a norm of u as the additional parameter. In general the norm of the anti-symmetric part u_a of u according to the splitting (2.8) with respect to a particular S may therefore be chosen

$$\begin{aligned} g(u,\lambda) &= 0, \\ \|u_a\|^2 - \rho^2 &= 0. \end{aligned} \tag{4.1}$$

The algorithm based on (4.1) is given by
For a given $\rho > 0$, $u^{(0)} \in X$, $\|u_a^{(0)}\| = \rho$, compute $(u^{(k)}, \lambda^{(k)})$ according to

$$u^{(k)} = \tilde{u}_s^{(k-1)} + \rho \frac{\tilde{u}_a^{(k-1)}}{\|\tilde{u}_a^{(k-1)}\|}, \quad u^{(k)*} g(u^{(k)}, \lambda^{(k)}) = 0,$$

$$\tilde{u}^{(k-1)} = u^{(k-1)} + \delta u^{(k-1)}, \quad k = 1, 2, \ldots, \text{ where} \tag{4.2}$$

$$\begin{bmatrix} g_u^{(k-1)} & g_\lambda^{(k-1)} \\ u_a^{(k-1)*} & 0 \end{bmatrix} \begin{bmatrix} \delta u^{(k-1)} \\ * \end{bmatrix} = - \begin{bmatrix} g^{(k-1)} \\ 0 \end{bmatrix}.$$

Here $u_a^{(k-1)*} \in (X')_a$ is such that $u_a^{(k-1)*} u_a^{(k-1)} = \|u_a^{(k-1)}\|$. Since $u_a/\rho \to \phi_0$ for $\rho \to 0$ along the non-symmetric branch from (2.11) we note that asymptotically the Jacobian of (4.2) is equal to that of the classical pseudo-arclength method. Therefore, the growth of the norm of its inverse is proportional to $|\rho-\rho_0|^{-1}$ on both branches as opposed to (3.3) (cf. [4]).

In the following we will prove a result relevant for the convergence of (4.2). For simplicity we consider the case $X = \mathbf{R}^N$. Let $I_{sa} = (I_s^T, I_a^T)^T$, where I_s, I_a are such that $I_{sa}\,\underline{u} = (\underline{u}_s^T, \underline{u}_a^T)^T$ yields a splitting of $\underline{u} \in \mathbf{R}^N$ into the symmetric and anti-symmetric components (cf. (5.4)).

Then (4.2) may be rewritten as

$$\begin{bmatrix} I_{sa} g_u^{(k-1)} I_{sa}^T & I_{sa} g_\lambda^{(k-1)} \\ \\ \frac{1}{\rho}(\underline{0}^T, \underline{u}_a^{(k-1)T}) & 0 \end{bmatrix} \begin{bmatrix} \delta\underline{u}_s^{(k-1)} \\ \delta\underline{u}_a^{(k-1)} \\ * \end{bmatrix} = - \begin{bmatrix} I_{sa} g^{(k-1)} \\ \\ 0 \end{bmatrix},$$

$$\underline{u}^{(k)} = I_{sa}^T \begin{bmatrix} \underline{u}_s^{(k-1)} + \delta\underline{u}_s^{(k-1)} \\ \\ \rho \dfrac{\underline{u}_a^{(k-1)} + \delta\underline{u}_a^{(k-1)}}{\|\underline{u}_a^{(k-1)} + \delta\underline{u}_a^{(k-1)}\|} \end{bmatrix}, \quad \underline{u}^{(k)T} g(\underline{u}^{(k)}, \lambda^{(k)}) = 0. \tag{4.3}$$

Now we write the result of one iteration in the fixed point form $\underline{u}^{(k)} = \Phi(\underline{u}^{(k-1)})$, where we consider λ to be a function of $\underline{u}$ through the Rayleigh-quotient in (4.3). Let $(\underline{u}_0, \lambda_0)$ be a solution of (4.1) and a regular point. Then define P_{u_0} to be the orthogonal projector onto the subspace $\{\underline{v} \in \mathbf{R}^N \,|\, \underline{v}_a \perp \underline{u}_{0a}\}$. If $\mathbf{R}^{N_s}$, $\mathbf{R}^{N_a}$ are the subspaces of the symmetric and antisymmetric parts, $N_s + N_a = N$, and I_{N_s}, I_{N_a} the identity matrices on these spaces then

$$P_{\underline{u}_0} = I_{sa}^T \begin{bmatrix} I_{N_s} & 0 \\ \\ 0 & I_{N_a} - \dfrac{\underline{u}_{0a}\underline{u}_{0a}^T}{\rho^2} \end{bmatrix} I_{sa}. \tag{4.4}$$

Proposition 4.1 Let $(\underline{u}_0, \lambda_0)$ be a regular solution of (4.1). The Jacobian Φ' of the mapping in (4.3) satisfies

$$\Phi'(\underline{u}_0) P_{\underline{u}_0} = 0. \tag{4.5}$$

Proof We denote the inverse of the matrix in (4.3) in $(\underline{u}_0, \lambda_0)$ by

$$\begin{bmatrix} H_0 & \underline{b}_0 \\ \underline{c}_0^T & d_0 \end{bmatrix} \quad , \quad H_0 = H(\underline{u}_0) \in \mathbf{R}^{N,N} . \tag{4.6}$$

Then

$$\Phi(\underline{u}) = I_{sa}^T \begin{bmatrix} I_{N_s} & 0 \\ & \\ 0 & \frac{\rho}{\|\underline{y}_a\|} I_{Na} \end{bmatrix} I_{sa} \underline{y}(\underline{u}),$$

where

$$\underline{y}(\underline{u}) = \underline{u} - I_{sa}^T H(\underline{u}) I_{sa} g(\underline{u}, \lambda(\underline{u}))$$

is the update of $\underline{u}$ provided by one Newton step for (4.1). In the solution $(\underline{u}_0, \lambda_0)$ we have

$$\underline{y}'(\underline{u}_0) I_{sa}^T = \frac{1}{\rho} I_{sa}^T \underline{b}_0 (\underline{0}^T, \underline{u}_{0a}^T) \tag{4.7}$$

from (4.3), (4.6).
Since it holds that

$$\Phi'(\underline{u}_0) = P_{\underline{u}_0} \underline{y}'(\underline{u}_0)$$

(4.5) is a consequence of (4.4) and (4.7).

5. A DUFFING EQUATION AND ITS DISCRETIZATION

From the theory of nonlinear vibrations the following problem is well-known (cf. [7,8,9]).

Let $X = C^{\infty}_{2\pi}(\mathbf{R})$ denote the space of 2π-periodic functions $u(t)$ that are infinitely differentiable on the real line.

$$\ddot{u} + 2u^3 = \lambda \cos t, \quad u \in X, \tag{5.1}$$

here $\ddot{u} = \frac{d^2}{dt^2} u$. Because of the symmetry properties of the cosine (2.6) is satisfied with $\gamma = (I, S_\alpha, S_\beta, S_\gamma)$, where

$$\begin{aligned} (S_\alpha u)(t) &= u(-t), \quad (S_\beta u)(t) = -u(\pi - t), \\ (S_\gamma u)(t) &= (S_\alpha S_\beta) u(t) = -u(t-\pi). \end{aligned} \tag{5.2}$$

There is a branch of solutions of (5.1) which are symmetric with respect to all symmetries and on this branch there are several bifurcation points at which two of the symmetries are broken and one is preserved.

In [9] the problem of directly computing these bifurcation points by using suitable extended systems was addressed. In order to obtain starting values for these methods a continuation up to a neighbourhood of the points is necessary. For this task the continuation methods proposed above may, of course, be used,but they also can be utilized to rapidly and accurately locate singular points (cf. [1]).

As in [9] we discretize (5.1) by a finite-difference method on an equidistant grid. Let $h = 2\pi/N$, $N = 4p$, p a positive integer. The gridpoints are then $t_i = ih - h/2$, $i = 1, \ldots, N$. Periodicity implies $u_i = u(t_i) = u_{N+i}$ on the extension of the grid covering the real line. The central difference operator for $\ddot{u}$ then yields the following finite-dimensional nonlinear eigenvalue problem

$$g(\underline{u}, \lambda) = A\underline{u} + \underline{F}(\underline{u}) - \lambda \underline{c} = 0,$$

$$\underline{u} = (u_1, \ldots, u_N)^T, \quad \underline{F} = (2u_1^3, \ldots, 2u_N^3)^T,$$

$$\underline{c} = (\cos t_1, \ldots, \cos t_N)^T, \tag{5.3}$$

$$A = \frac{1}{h^2} \begin{bmatrix} -2 & 1 & 0 & \ldots & 0 & 0 & 1 \\ 1 & -2 & 1 & 0 \ldots & 0 & 0 & 0 \\ 0 & 1 & -2 & 0 \ldots & 0 & 0 & 1 \\ & & & \vdots & & & \\ 1 & 0 & 0 & 0 \ldots & 0 & 1 & -2 \end{bmatrix} \in \mathbf{R}^{N,N}.$$

Using the unit vectors $\underline{e}_i \in \mathbf{R}^N$, $i = 1, \ldots, N$, we can give basis representations for these spaces

$$X_s = \text{span}\,\{\underline{e}_i - \underline{e}_{2p+1-i} - \underline{e}_{2p+i} + \underline{e}_{N+1-i},\ i = 1, \ldots, p\},$$

$$X_a^1 = \text{span}\,\{\underline{e}_i + \underline{e}_{2p+1-i} + \underline{e}_{2p+i} + \underline{e}_{N+1-i},\ i = 1, \ldots, p\},$$

$$X_a^2 = \text{span}\,\{\underline{e}_i - \underline{e}_{2p+1-i} + \underline{e}_{2p+i} - \underline{e}_{N+1-i},\ i = 1, \ldots, p\},$$

$$X_a^3 = \text{span}\,\{\underline{e}_i + \underline{e}_{2p+1-i} - \underline{e}_{2p+i} - \underline{e}_{N+1-i},\ i = 1, \ldots, p\}.$$

Now $\underline{x} \in \mathbf{R}^N$ may be decomposed into its symmetric (w.r.t. all three symmetries) and "anti-symmetric" (w.r.t. two of the symmetries) parts

$$\underline{x}_s = I_s \underline{x}, \qquad \underline{x}_a^i = I_a^i \underline{x}, \qquad i = 1,2,3,$$

where $I_s, I_a^i : \mathbf{R}^N \to \mathbf{R}^p$, $i = 1,2,3$, are given by

$$I_s = \frac{1}{2}\,[I_p, -E_p, -I_p, E_p],$$

$$I_a^1 = \frac{1}{2}\,[I_p, E_p, I_p, E_p],$$

$$I_a^2 = \frac{1}{2}\,[I_p, -E_p, I_p, -E_p],$$

$$I_s^3 = \frac{1}{2}\,[I_p, E_p, -I_p, -E_p].$$

We then have $I_{sa}\underline{x} = (\underline{x}_s^T, \underline{x}_a^{1T}, \underline{x}_a^{2T}, \underline{x}_a^{3T})^T$ for

$$I_{sa} = [I_s^T, I_a^{1T}, I_a^{2T}, I_a^{3T}]^T. \tag{5.4}$$

6. NUMERICAL IMPLEMENTATION

The definitions of the previous section allow us to formulate several algorithms to solve specific parts of the problem (5.1). To compute the symmetric (w.r.t. all 3 symmetries) branch only the problem

$$I_s g(I_s^T \underline{u}_s, \lambda) = 0 \tag{6.1}$$

may be solved by (1.3). It is p-dimensional in p unknowns and there are no symmetry-breaking bifurcation points along the branch.

In order to find the symmetry-breaking bifurcation points and the solutions branching off at those we solve the full problem in the form

$$I_{sa} g(I_{sa}^T \underline{u}, \lambda) = 0. \tag{6.2}$$

If a bifurcation point (u_0, λ_0) is encountered and an approximate null vector $\underline{\phi}_0$ of the Jacobian g_u^0 is computed then the decomposition $I_{sa}\underline{\phi}_0$ will show which symmetries are broken on the bifurcating branch. Say, the symmetries S_α and S_β are broken while S_γ is preserved, then $I_a^3 \underline{\phi}_0 \neq 0$ and the following variant of algorithm (4.2) may be used to compute solutions on the non-symmetric branch.

$$\begin{bmatrix} I_{sa} g_u^{(k-1)} I_{sa}^T & I_{sa} g_\lambda^{(k-1)} \\ & \\ & \\ \frac{1}{\rho}(\underline{0}^T, \underline{0}^T, \underline{0}^T, \underline{u}^{(k-1)T} I_a^{3T}) & 0 \end{bmatrix} \begin{bmatrix} \delta\underline{u}_s^{(k-1)} \\ \delta\underline{u}_a^{1(k-1)} \\ \delta\underline{u}_a^{2(k-1)} \\ \delta\underline{u}_a^{3(k-1)} \end{bmatrix} = \begin{bmatrix} -I_{sa} g^{(k-1)} \\ \\ \\ 0 \end{bmatrix}$$

$$\underline{u}^{(k)} = I_{sa}^T \begin{bmatrix} \underline{u}_s^{(k-1)} + \delta\underline{u}_s^{(k-1)} \\ \underline{u}_a^{1(k-1)} + \delta\underline{u}_a^{1(k-1)} \\ \underline{u}_a^{2(k-1)} + \delta\underline{u}_a^{2(k-1)} \\ \rho \dfrac{\underline{u}_a^{3(k-1)} + \delta\underline{u}_a^{3(k-1)}}{\|\underline{u}_a^{3(k-1)} + \delta\underline{u}_a^{3(k-1)}\|} \end{bmatrix}, \quad \lambda^{(k)} = \frac{\underline{c}^T(A\underline{u}^{(k)} + \underline{F}(\underline{u}^{(k)}))}{\underline{c}^T\underline{c}},$$

where $\rho > 0$, $u^{(0)} = \underline{u}_0 + \rho \frac{I_a^3\phi_0}{\| I_a^3\phi_0 \|}$, $\lambda^{(0)} = \frac{\underline{c}^T(A\underline{u}^{(0)} + \underline{F}(\underline{u}^{(0)}))}{\underline{c}^T\underline{c}}$.

We note that for (5.3) the Rayleigh-quotient in (4.2) may explicitely be solved for $\lambda^{(k)}$ and no iterative solution is necessary.

7. NUMERICAL RESULTS

The discretization (5.3) of the problem (5.1) was solved using algorithm (6.3). The branch of solutions that possess all three symmetries S_1, S_2 and S_3 exhibits a series of bifurcation points at which two of these symmetries are broken for all possible combinations of those ([7, 9]).

We present below the results for the first three of these bifurcation points encountered along the branch bifurcating at the origin. As described in the previous section a bifurcation point and the symmetries broken there were detected and yielded starting values for the numerical solution. Computation of the non-symmetric solutions on the bifurcating branches required an average of 2-3 corrector iterations independent of N and for relatively large steps along the branch.

Figure 7.1a shows the primary branch marked by "S" and the three bifurcating branches

"A" S_1 and S_2 - symmetry broken,
"B" S_1 and S_3 - symmetry broken,
"C" S_2 and S_3 - symmetry broken.

Since on a ρ-λ diagram branch A cannot be easily distinguished we have added Figure 7.1b in which the solution at the node $t_1 = \pi/N$ is depicted versus λ. Figures 7.2 - 7.5 show the solutions along the primary branch and the branches A, B and C for various points along each branch.

REFERENCES

[1] R. E. Bank, PLTMG User's Guide, Edition 4.0, Tech. Report, Dept. Math., University of California, San Diego (1985).

[2] D. W. Decker and H. B. Keller, Path following near bifurcation, Comm. Pure Appl. Math. 34, 149-175 (1981).

[3] H. D. Mittelmann, An efficient algorithm for bifurcation problems of variational inequalities, Math. Comp. 41, 473-485 (1983).

[4] H. D. Mittelmann, Continuation near symmetry-breaking bifurcation points, in Numerical Methods for Bifurcation Problems (T. Küpper, H. D. Mittelmann and H. Weber, eds.), ISNM 70, Birkhäuser - Verlag, Basel, 1984.

[5] H. D. Mittelmann, Multi-level continuation techniques for nonlinear problems with parameter dependence, Appl. Math. Comp. (to appear).

[6] G. H. Pimbley, Eigenfunction branches of non-linear operators and their bifurcation, Lecture Notes in Mathematics, vol. 104, Springer-Verlag, Berlin, 1969.

[7] B. V. Schmitt, Sur la structure de l'equation de Duffing sans Dissipation. SIAM J. Appl. Math. 42, 868-883 (1982).

[8] R. Seydel, Branch switching in bifurcation problems for ordinary differential equations, Numer. Math. 41, 93-116 (1983).

[9] B. Werner and A. Spence, The computation of symmetry-breaking bifurcation points, SIAM J. Numer. Anal. 21, 388-399 (1984).

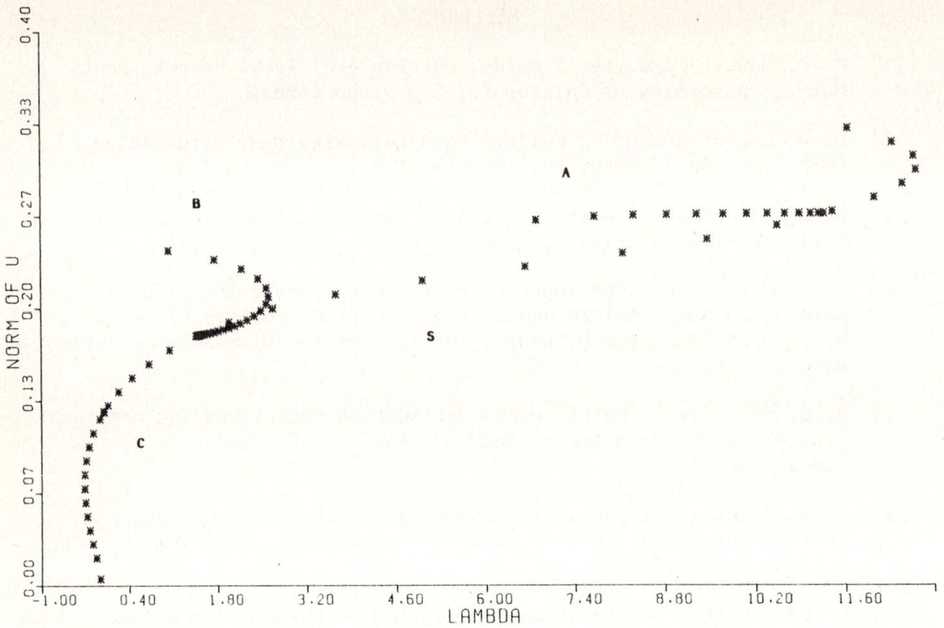

Fig. 7.1a Symmetric branch and first three symmetry-breaking branches; ρ-λ diagram.

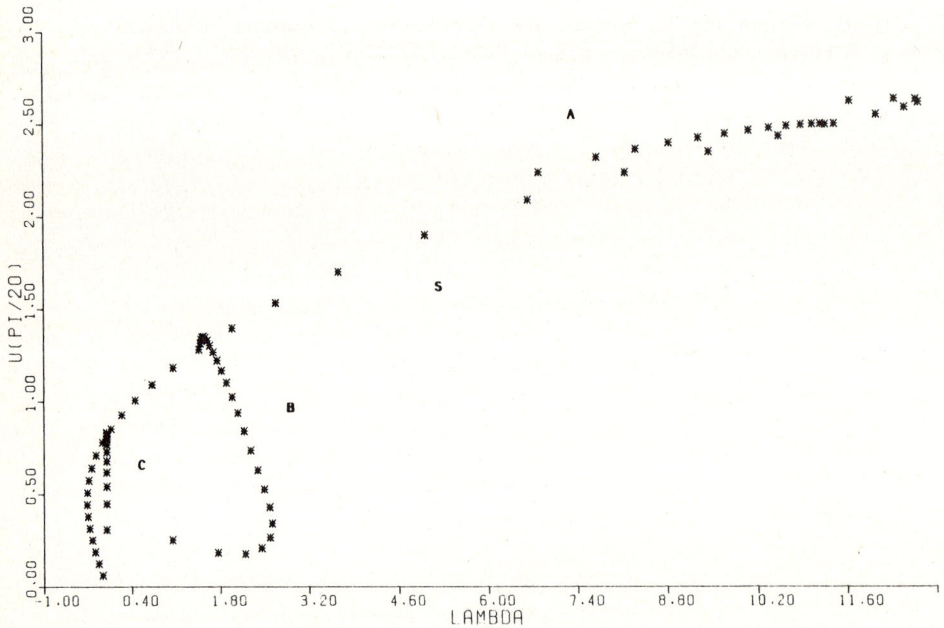

Fig. 7.1b Symmetric branch and first three symmetry-breaking branches; $\underline{u}_1$-λ diagram.

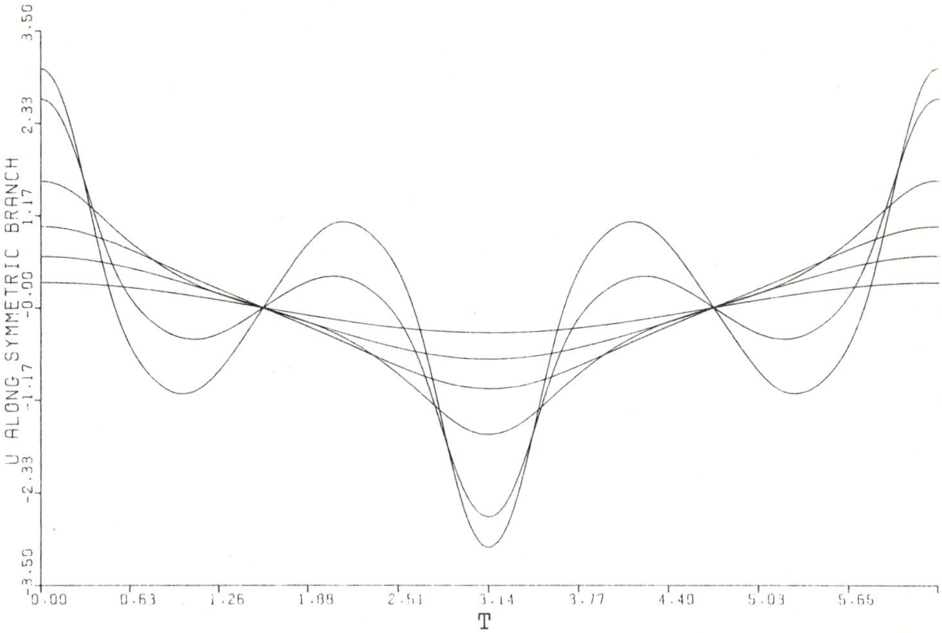

Fig. 7.2 Solution on symmetric branch.

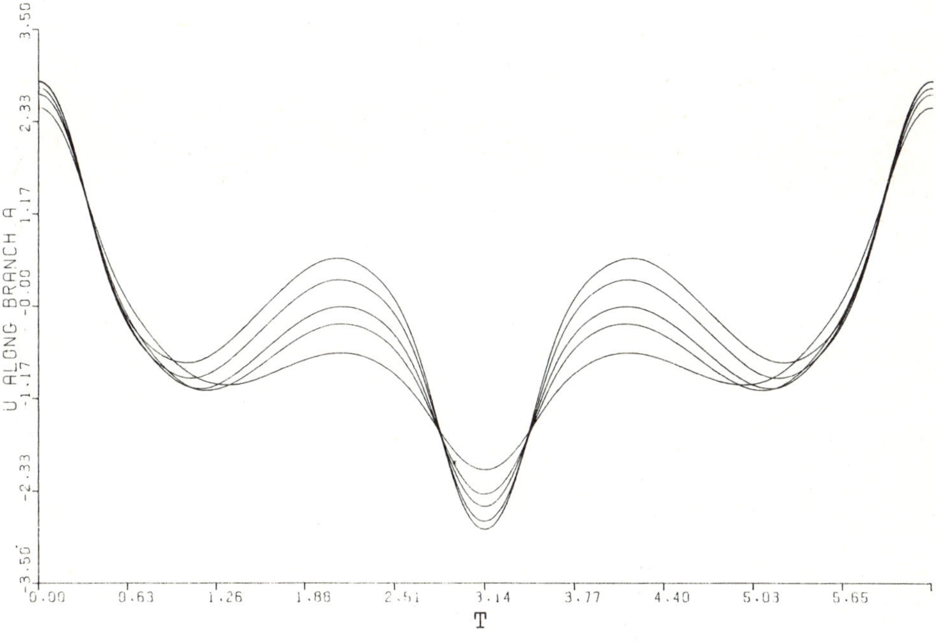

Fig. 7.3 Solution on S_1 and S_2 -breaking branch.

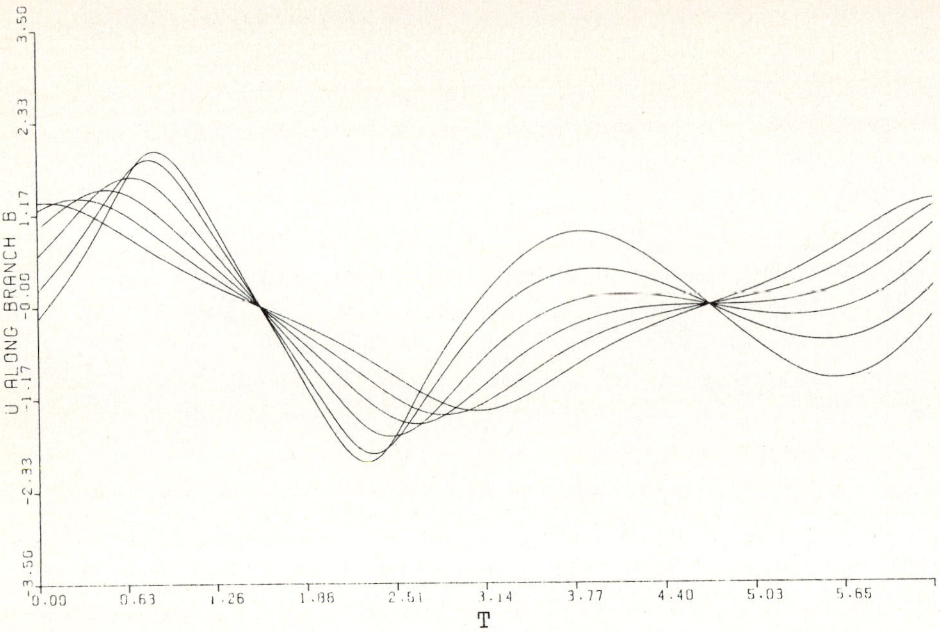

Fig. 7.4 Solution on S_1 and S_3 -breaking branch.

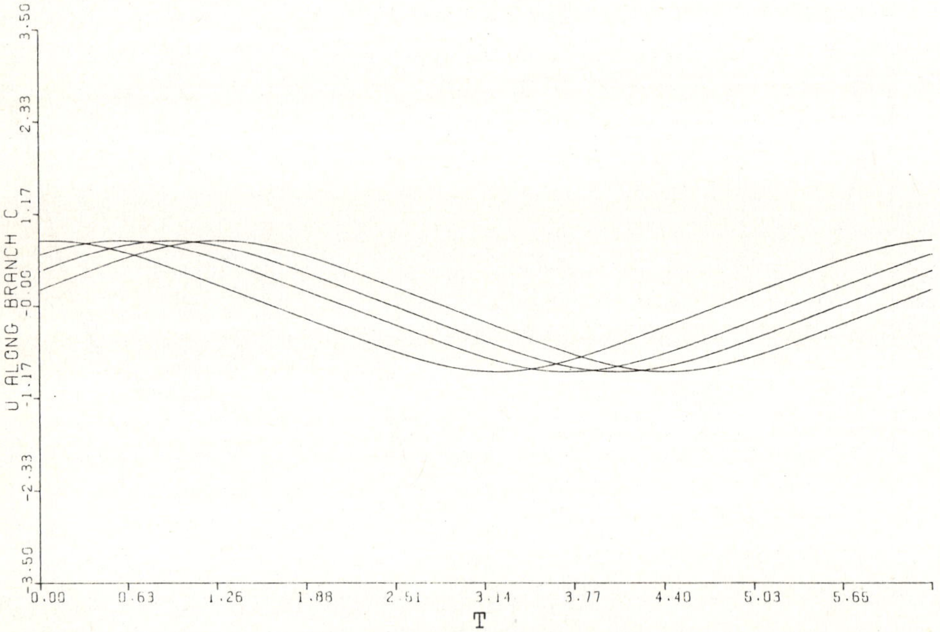

Fig. 7.5 Solution on S_2 and S_3 -breaking branch.

POSTPROCESSING OF A FINITE ELEMENT SCHEME WITH LINEAR ELEMENTS

P. Neittaanmäki[1] and M. Křížek[2]

[1] University of Jyväskylä, Department of Mathematics, Seminaarinkatu 15, SF-40100 Jyväskylä, Finland

[2] Czechoslovak Academy of Sciences, Mathematical Institute, Žitná 25, CS-11567 Prague 1, Czechoslovakia

SUMMARY

In this contribution we first give a brief survey of postprocessing techniques for accelerating the convergence of finite element schemes for elliptic problems. We also generalize a local superconvergence technique recently analyzed by Křížek and Neittaanmäki ([20]) to a global technique. Finally, we show that it is possible to obtain $O(h^4)$ accuracy for the gradient in some cases when only linear elements are used. Numerical tests are presented.

1. INTRODUCTION

Let us consider a general FE-scheme illustrated in Figure 1.1.

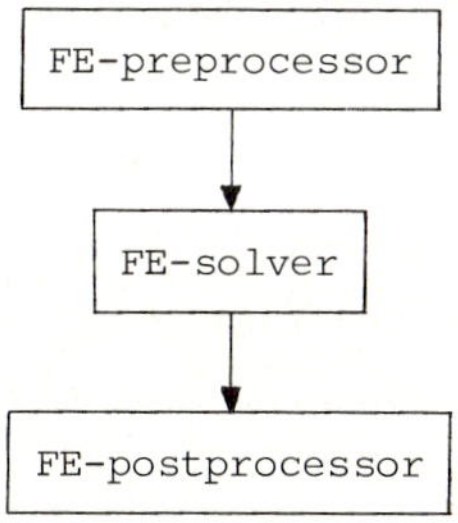

Figure 1.1. FE-scheme

First of all we briefly outline recent trends in this field of investigation.

In connection with the FE-preprocessor, much work has been done for finding an "optimal" position of a FE-grid, and for splitting the FE-scheme (domain decomposition methods, substructuring) in order to use parallel processors, etc., see [10, 11, 12] and references given therein.

The FE-solver consists of routines for solving linear or nonlinear systems of equations, or corresponding eigenvalue problems. For a review of such routines (relaxation methods, conjugate gradient methods, preconditioning, incomplete decomposition, etc.) we refer to the monograph [3]. During the last ten years a lot of attention has been focused also to defect correction and multigrid methods, see the monograph [14].

The postprocessing of the FE-scheme seems to be a very popular and efficient tool. It may recover not only FE-solution, but also its derivatives. There are many different approaches. One can use several averaging techniques, integral (convolution) smoothing techniques or solve an auxiliary problem to increase accuracy. Concerning postprocessing procedures for mixed finite element methods we refer to [6, 35]. Recently developed extrapolation methods [5, 33] are a promising tool as well. This paper presents several simple postprocessing techniques.

In Section 2 we give a short survey of superconvergence techniques and especially techniques for improving the accuracy of the gradient when linear elements are employed to the problem

$$\begin{cases} -\Delta u = f & \text{in } \Omega , \\ u = 0 & \text{on } \partial\Omega . \end{cases} \qquad (1.1)$$

Here $\Omega \subset R^2$ is a bounded domain and $f \in L^2(\Omega)$. For the sake of clarity, we demonstrate the superconvergence schemes in their simplest setting.

In Section 3 we generalize a local superconvergence technique we have recently analyzed [20] for an elliptic problem, to a global technique. We recall the main idea of the proposed scheme. The use of linear elements yields a piecewise constant field of the first derivatives of the finite element solution. Thus employing a suitable postprocessing (based on averaging at nodes), a new continuous piecewise linear field may be determined, which has better approximation properties.

Section 4 is devoted to another technique which yields even the local $O(h^4)$-superconvergence of a postprocessed gradient for linear elements.

2. A SHORT SURVEY OF TECHNIQUES YIELDING SUPERCONVERGENCE

2.1. Preliminaries

Let us introduce some notations and definitions.

Let $\Omega \subset \mathbb{R}^2$ be a bounded domain with a Lipschitz boundary $\partial\Omega$. We write $\Omega_0 \subset\subset \Omega$ if Ω_0 is a domain such that $\bar{\Omega}_0 \subset \Omega$. By $P_k(\Omega)$ we mean the space of polynomials of the degree at most k.

The usual norm and seminorm in the Sobolev space $(W_p^k(\Omega))^d$ ($k \in \{1,2,\dots\}$), $d \in \{1,2\}$, $p \in [1,\infty]$) are denoted by $\|\cdot\|_{k,p,\Omega}$ and $|\cdot|_{k,p,\Omega}$, respectively.

In particular, we write $H^k(\Omega) = W_p^k(\Omega)$ and $\|\cdot\|_k = \|\cdot\|_{k,\Omega} = \|\cdot\|_{k,p,\Omega}$ for $p = 2$. Let $(L^2(\Omega))^d = (H^0(\Omega))^d$ be equipped with the scalar product $(\cdot,\cdot)_{0,\Omega}$.

The space $H_0^1(\Omega)$ is a closure of $C_0^\infty(\Omega)$ under the norm $\|\cdot\|_{1,\Omega}$; i.e.

$$H_0^1(\Omega) = \{\varphi \in H^1(\Omega) \mid \varphi = 0 \text{ on } \partial\Omega\}.$$

For further purpose we recall the variational formulation of (1.1): Find $u \in H_0^1(\Omega)$ such that

$$(\nabla u, \nabla\varphi)_{0,\Omega} = (f,\varphi)_{0,\Omega} \qquad \forall\, \varphi \in H_0^1(\Omega). \tag{2.1}$$

In what follows the letter C stands for a generic positive constant which may vary with context.

2.2. Error estimates for the linear elements

Let $\{T_h\}$ be a regular family of triangulations of $\bar{\Omega}$ (i.e. the well-known Zlámal's condition on the minimal angle is fulfilled). We shall consider the Galerkin approximation of (2.1) which consists in finding $u_h \in V_h$ such that

$$(\nabla u_h, \nabla v_h)_{0,\Omega} = (f, v_h)_{0,\Omega} \qquad \forall\, v_h \in V_h, \tag{2.2}$$

where V_h is the space of piecewise linear trial functions:

$$V_h = \{v_h \in C(\bar{\Omega}) \mid v_h = 0 \text{ on } \partial\Omega,\ v_h|_T \in P_1(T) \quad \forall\, T \in T_h\}.$$

We recall the following optimal error estiamtes [34, 13]:

$$\|u - u_h\|_{0,p,\Omega} = \begin{cases} C_p h^2 \|u\|_{2,p,\Omega} & \text{if } p \in [2,\infty), \\ Ch^2 |\ln h|\ \|u\|_{2,\infty,\Omega} & \text{if } p = \infty, \end{cases} \tag{2.3}$$

$$\|\nabla u - \nabla u_h\|_{0,p,\Omega} \leq Ch \|u\|_{2,p,\Omega} \qquad \text{if} \quad p \in [2,\infty] . \tag{2.4}$$

We point out that all error estimates considered here are of an asymptotic nature, i.e. all assertions will hold only for a sufficiently small triangulation parameter.

Note that (see [5, 23]) the logarithm factor in (2.3) does not occur when triangulations $\mathcal{T}_h$ are uniform; such triangulations are defined by the property that any two adjacent (closed) triangles of $\mathcal{T}_h$ form a parallelogram. Moreover, if each $\mathcal{T}_h$ consists of equilateral triangles then we have the superconvergence estimate [5, 28]

$$\max_{x \in N_h} |u(x) - u_h(x)| \leq Ch^4 \|u\|_{C^{4+\varepsilon}(\bar{\Omega})} \qquad (\varepsilon > 0) , \tag{2.5}$$

where N_h is the set of all nodes of $\mathcal{T}_h$. In [33] this result is extended to elasticity problems.

2.3. Superconvergent recovery of the gradient

The $O(h^2)$-superconvergence of the tangential component of the gradient at midpoints of sides of uniform (quasiuniform) triangulations has been established by [2, 24, 25]. Denoting the set of these midpoints by M_h, one can recover both the components of the gradient employing the following postprocessing (firstly proposed in [7])

$$\tilde{\nabla} u_h(x) = \frac{1}{2}(\nabla u_h|_{T_1} + \nabla u_h|_{T_2}) , \qquad x \in M_h \cap \Omega , \tag{2.6}$$

where $T_1, T_2 \in \mathcal{T}_h$ are those adjacent triangles for which the midpoint $x \in T_1 \cap T_2$. For the recovered gradient $\tilde{\nabla} u_h$ it holds that (cf. (2.4))

$$\max_{x \in M_h \cap \Omega} \|\nabla u(x) - \tilde{\nabla} u_h(x)\|_{R^2} \leq Ch^2 |\ln h| \, \|u\|_{3,\infty,\Omega} \tag{2.7}$$

or even $O(h^2)$ for the discrete L^2-norm [24, 25]. An analogous superconvergence estimate has been later derived even for a class of nonlinear problems [8].

Let $T_1, T_2, T_3 \in \mathcal{T}_h$ be the adjacent triangles of $T \in \mathcal{T}_h$, $T \subset \Omega_0 \subset\subset \Omega$. Then using (2.6) one can easily derive [25] the scheme

$$\tilde{\nabla} u_h(x) = \frac{1}{6}(3\nabla u_h|_T + \nabla u_h|_{T_1} + \nabla u_h|_{T_2} + \nabla u_h|_{T_3}) , \quad x \in C_h \cap \Omega_0 \tag{2.8}$$

which yields the superconvergence of the gradient at the centroids $x \in C_h$ of triangles in Ω_0. With the help of (2.6), we may define a discontinuous piecewise linear field $\widetilde{\nabla u}_h$ which recovers the gradient at any point of $\Omega_0 \subset\subset \Omega$ (see [31]).

Averaging at nodes may also lead to superconvergence of the gradient. Putting

$$\widetilde{\nabla u}_h(x) = \frac{1}{6} \sum_{T \cap \{x\} \neq \emptyset} \nabla u_h|_T , \qquad x \in N_h \cap \Omega , \tag{2.9}$$

on uniform meshes and using further a convenient averaging formula (see (3.15)) at boundary nodes, we can determine a continuous piecewise linear field $\widetilde{\nabla u}_h$ on the whole $\bar{\Omega}$ to obtain the global superconvergence error estimate ([22])

$$\| \nabla u - \widetilde{\nabla u}_h \|_{0,p,\Omega} \leq Ch^2 |\ln h|^{1-2/p} \|u\|_{3,p,\Omega} , \quad p \in \{2,\infty\} . \tag{2.10}$$

Employing the scheme (2.9) to problems where $\partial\Omega$ is smooth (T_h are not uniform near the boundary $\partial\Omega$), we have proved [20] the local $O(h^{3/2})$-superconvergence of the gradient in the $(L^2(\Omega_0))^2$-norm ($\Omega_0 \subset\subset \Omega$). In Section 3 of this paper we generalize this result to obtain a global superconvergence of the same order.

A wide class of postprocessings, where $\Omega_0 \subset\subset \Omega$ is covered by uniform triangulations with right-angled isosceles triangles, is introduced in [37] (see also [36] for the finite difference method). For instance, setting

$$(\widetilde{\nabla u}_h(x))_1 = \frac{1}{2h}(u_h(x_1 + h, x_2) - u_h(x_1 - h, x_2)) , \tag{2.11}$$

$$x = (x_1, x_2) \in N_h \cap \Omega_0$$

(the second component $(.)_2$ is defined analogically), we get ([37], p. 81)

$$\max_{x \in N_h \cap \Omega_0} \| \nabla u(x) - \widetilde{\nabla u}_h(x) \|_{R^2} \leq Ch^2 (\|u\|_{5,\infty,\Omega_1} + \|u - u_h\|_{L^1(\Omega_1)}) \tag{2.12}$$

where $\Omega_0 \subset\subset \Omega_1 \subset\subset \Omega$. In Section 4, we present another type of such a postprocessing which yields even the nodal $O(h^4)$-superconvergence of the gradient in Ω_0 for a special second order problem.

For the same type of triangulations as in the foregoing case, another type of postprocessing operator is presented in [32], p. 148,

$$\tilde{u}_h(x) = \frac{1}{4h^2} \int_{D_h} u_h(x + y)\, dy , \qquad x \in \Omega_0 \subset\subset \Omega ,$$

where $D_h = (-h,h) \times (-h,h)$. If $\partial\Omega$ is smooth it holds that

$$\|u - \tilde{u}_h\|_{1,\Omega_0} \leq Ch^{3/2} \|u\|_{3,\Omega} ,$$

which is, in fact, a superconvergent estimate for the gradient.

Further type of integral smoothing operator is given in [4]. It is applicable even on irregular meshes, but it is not local as all previous schemes, i.e. more arithmetical operations have to be done. In [16] a local and global least squares smoothing of ∇u_h is proposed to obtain a better approximation to ∇u. Other superconvergence phenomena with linear elements are described in [15, 27, 29, 35], see also survey papers [9, 21, 26].

3. GLOBAL SUPERCONVERGENCE OF A POSTPROCESSED GRADIENT

The H^3-regularity of the solution u in polygonal domains (see Section 2) can be verified apriori only in some special cases (see [19]). For instance, if Ω is an acute-angled triangle and $f \in H^1(\Omega)$, or if Ω is a rectangle and f belongs to a weighted Sobolev space, then the weak solution u of (1.1) is in $H^3(\Omega)$. In this section we firstly deal with domains with smooth boundaries, where the H^3-regularity is more acceptable. However, we shall be not able to derive the $O(h^2)$-superconvergence of the gradient like in (2.10), but only $O(h^{3/2})$, since triangulations near boundary will be irregular.

Consider the problem

$$\begin{cases} -\text{div}(A \,\text{grad}\, u) + au = f & \text{in } \Omega , \\ \qquad\qquad\qquad\qquad u = 0 & \text{on } \partial\Omega , \end{cases} \tag{3.1}$$

where $f \in L^2(\Omega)$, $a \in R^1$, $a \geq 0$, $A \in (C^2(\bar{\Omega}))^4$ is a symmetric uniformly positive definite 2×2 matrix and $\Omega \subset \mathbb{R}^2$ is a bounded domain with a Lipschitz boundary $\partial\Omega$, which is supposed to be three-times continuously differentiable and let $\partial\Omega$ consist of a finite number of convex and concave arcs. For such a domain, for $v \in H^1(\Omega)$ and for sufficiently small $\varepsilon > 0$, it holds that (see [18], Th. 4.4 and Rem. 12.4)

$$\|v\|_{0,\Omega^\varepsilon} \leq C\varepsilon^{1/2} \|v\|_{1,\Omega} , \tag{3.2}$$

where $\Omega^\varepsilon \subset \Omega$ is a boundary strip of the width ε. We use this inequality in the proof of Theorem 3.1 to estimate derivatives of u near $\partial\Omega$.

There are polygonal approximations Ω_h of Ω such that $\Omega_h \subset \Omega$ and

$$\max_{x \in \partial\Omega} \operatorname{dist}(x, \partial\Omega_h) \leq Ch^2 \tag{3.3}$$

We assume that the boundary $\partial\Omega_h$ consists of chords or tangents of convex or concave arcs, respectively, and that any point of inflexion of $\partial\Omega$ coincide with a vertex of $\bar{\Omega}_h$.

Let $\{T_h\}$ be a regular family of triangulations of $\bar{\Omega}_h$, and let the ratio of the lengths of any two sides in any T_h is not less than a given constant $C \geq 1$ (i.e. $\{T_h\}$ is strongly regular). Moreover, let $T_h^* \subset T_h$ be a uniform triangulation which is "maximal" in the following sense

$$\max_{x \in \partial\Omega} \operatorname{dist}(x, \partial\Omega_h^*) \leq Ch ,$$

where (see Figure 3.1)

$$\Omega_h^* = \bigcup_{T \in T_h^*} T .$$

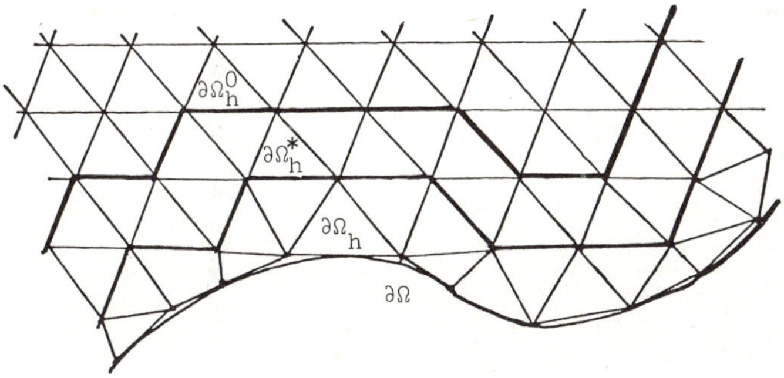

Figure 3.1.

By u_h we denote the Galerkin approximation of u in the space

$$V_h = \{v_h \in C(\bar{\Omega}) \mid v = 0 \text{ on } \Omega - \Omega_h ,\ v_h|_T \in P_1(T) \quad \forall\, T \in T_h\} ,$$

and by $N_h^0 \subset N_h$ we mean the set of all internal nodes of T_h^*. Setting

$$\bar{\Omega}_h^0 = \bigcup\{T \in T_h \mid \exists\, x \in N_h^0 : T \cap \{x\} \neq \emptyset\} ,$$

we see that $\Omega_h^0 \subset \Omega_h^*$ and that

$$\max_{x \in \partial\Omega} \operatorname{dist}(x, \partial\Omega_h^0) \leq Ch . \tag{3.4}$$

The averaged gradient in $\bar{\Omega}_h^0$ will be a continuous piecewise linear field which is determined by (2.9), and we put

$$\widetilde{\nabla} u_h = \nabla u_h \quad \text{in} \quad \Omega_h^1 = \Omega_h - \bar{\Omega}_h^0 .$$

Theorem 3.1. If $f \in H^1(\Omega)$ then for sufficiently small h it is

$$\|\nabla u - \widetilde{\nabla} u_h\|_{0,\Omega_h} \leq C h^{3/2} \|f\|_{1,\Omega} , \tag{3.5}$$

where u is the weak solution of (3.1) and $u_h \in V_h$ its Galerkin approximation.

Proof. First of all we estimate the term

$$\|\nabla u - \widetilde{\nabla} u_h\|_{0,\Omega_h^1} = \|\nabla u - \nabla u_h\|_{0,\Omega_h^1} \leq \|u - u_h\|_{1,\Omega_h^1} \tag{3.6}$$

$$\leq \|u - P_h u\|_{1,\Omega_h^1} + \|P_h u - u_h\|_{1,\Omega} .$$

Here $P_h u \in V_h$ is an interpolation of u defined in this manner:

$$P_h u(x) = u(x) \qquad \forall\, x \in N_h - \partial\Omega_h .$$

Note that u is continuous as $H^3(\Omega) \subset C(\bar{\Omega})$ and (see [1])

$$\|u\|_{3,\Omega} \leq C \|f\|_{1,\Omega} .$$

Further we introduce the standard linear interpolation of u which we define only on Ω_h

$$\pi_h u(x) = u(x) \qquad \forall\, x \in N_h .$$

Let $x \in N_h$ be such that $x \in \partial\Omega_h$. Then by (3.3)

$$|\pi_h u(x) - P_h u(x)| = |u(x) - 0| \leq C h^2 \|\nabla u\|_{C(\bar{\Omega})} \leq C' h^2 \|u\|_{3,\Omega} .$$

Hence, for any $T \in \mathcal{T}_h$, $T \subset \Omega_h^1$, we have

$$\|u - P_h u\|_{1,T} \leq \|u - \pi_h u\|_{1,T} + \|\pi_h u - P_h u\|_{1,T} \tag{3.7}$$

$$\leq C h \|u\|_{2,T} + C'' h^2 \|u\|_{3,\Omega} .$$

Due to the strong regularity of the family $\{\mathcal{T}_h\}$, the number of triangles in Ω_h^1 is proportional to h^{-1}. Consequently,

$$\|u - P_h u\|_{1,\Omega_h^1} \leq C h \|u\|_{2,\Omega_h^1} + C''' h^{3/2} \|u\|_{3,\Omega} . \tag{3.8}$$

Applying now the Iljin inequality (3.2), we get by (3.4) and (3.8)

$$\|u - P_h u\|_{1,\Omega_h^1} \leq C h^{3/2} \|u\|_{3,\Omega} . \tag{3.9}$$

According to [17], Part I, Th. 2.1 (or [32]) it is

$$\|P_h u - u_h\|_{1,\Omega} \le Ch^{3/2} \|u\|_{3,\Omega} . \tag{3.10}$$

Substituting this and (3.9) into (3.6), we come to

$$\|\nabla u - \tilde{\nabla} u_h\|_{0,\Omega_h^1} \le Ch^{3/2} \|u\|_{3,\Omega} . \tag{3.11}$$

Further we deal with the term

$$\|\nabla u - \tilde{\nabla} u_h\|_{0,\Omega_h^0} \le \|\nabla u - \tilde{\nabla}(P_h u)\|_{0,\Omega_h^0} + \|\tilde{\nabla}(P_h u - u_h)\|_{0,\Omega_h^0} . \tag{3.12}$$

Let $\bar{u} \in H^3(R^2)$ be the Calderon extension of the function u on Ω (see [30], p. 80), i.e. $\bar{u}|_\Omega = u$ and

$$\|\bar{u}\|_{3,R^2} \le C\|u\|_{3,\Omega} . \tag{3.13}$$

We define a linear interpolation $\bar{P}_h \bar{u}$ of $\bar{u}$ over the whole $\mathbb{R}^2$, which is covered by uniform triangulations, so that $\bar{P}_h \bar{u} = P_h u$ in Ω_h^0. Let $\tilde{\nabla}(\bar{P}_h \bar{u})$ be defined in the whole plane R^2 with the help of the averaging formula (2.9). Then by [20], p. 111, and by (3.13) we have

$$\begin{aligned}\|\nabla u - \tilde{\nabla}(P_h u)\|_{0,\Omega_h^0} &\le \|\nabla\bar{u} - \tilde{\nabla}(\bar{P}_h \bar{u})\|_{0,R^2} \le Ch^2 \|\bar{u}\|_{3,R^2} \\ &\le C' h^2 \|u\|_{3,\Omega} .\end{aligned} \tag{3.14}$$

Moreover, according to [20], p. 113, and (3.10)

$$\|\tilde{\nabla}(P_h u - u_h)\|_{0,\Omega_h^0} \le \sqrt{13}\, \|\nabla(P_h u - u_h)\|_{0,\Omega} \le Ch^{3/2} \|u\|_{3,\Omega} .$$

Therefore, by (3.14) and (3.12)

$$\|\nabla u - \tilde{\nabla} u_h\|_{0,\Omega_h^0} \le Ch^{3/2} \|u\|_{3,\Omega} .$$

This together with (3.11), (3.6) and (3.7) yields the assertion (3.5). □

Suppose now that Ω is a polygon which is covered by uniform triangulations. In [21, 22], we propose a convenient choice of averaging formula at boundary nodes (see (3.15) below) which yields (2.10). The superconvergence estimate (2.10), for $p = 2$, is generalized in [17] also to elliptic systems. We shall illustrate now this technique in the following test example.

<u>Example 3.1.</u> Let $\Omega = (0,1) \times (0,1)$ and let the triangulations of

Ω be uniform (see Figure 3.2).

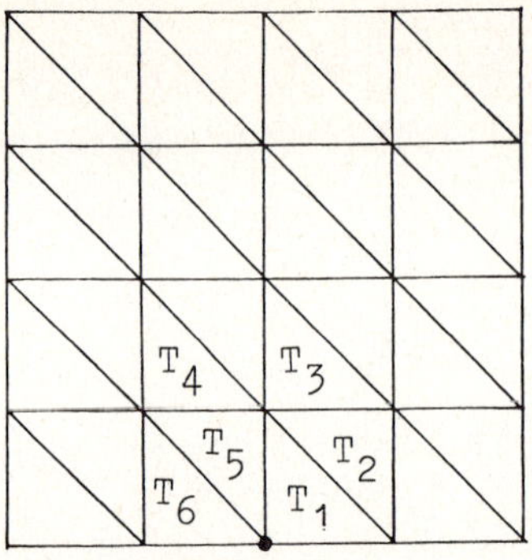

Figure 3.2.

We choose the averaged gradient as follows:

$$\tilde{\nabla} u_h(x) = \begin{cases} \frac{1}{6} \sum_{T \cap \{x\} \neq \emptyset} \operatorname{grad} u_h|_T , & x \in N_h \cap \Omega \\ \sum_{i=1}^{6} w_i \operatorname{grad} u_h|_{T_i} , & x \in N_h \cap (\partial\Omega_h \setminus Y) \\ 0 , \quad x \in Y , \end{cases} \tag{3.15}$$

where Y is the set of vertices of $\bar{\Omega}$ and T_i are indicated in Figure 3.2. The weights w_i are chosen so that $w_1 + \ldots + w_6 = 1$. In the following test we have (for numbering we refer to Figure 3.2) $w_1 = 3/4$, $w_2 = 0$ $w_3 = -1/4$, $w_4 = -1/4$, $w_5 = 0$, $w_6 = 3/4$. For further discussion of an other choice of weights in (3.15) we refer to ([22, 25, 17]).

For $f = 2 \sin \pi x_2 + \pi^2 x_1 (1 - x_1) \sin \pi x_2$ the exact solution of (1.1) is $u = x_1 (1 - x_1) \sin \pi x_2$.

In Table 3.2 we see a comparison of the L^2-accuracy of the gradients obtained from (2.2) by standard numerical derivation ∇u_h and by averaging technique (3.15) $\tilde{\nabla} u_h$

Table 3.1. Comparison of accuracies

h	$\|\nabla u - \nabla u_h\|_{0,\Omega}$	$\|\nabla u - \tilde{\nabla} u_h\|_{0,\Omega}$
1/4	.3324	.1343
1/8	.1741	.0396
1/16	.0881	.0106
1/32	.0441	.0027
rate of convergence	$0(h)$	$0(h^2)$

We find that the averaging technique gives the rate $O(h^2)$. The growth of the CPU time due to the postprocessing technique is essentially negligible. Typical CPU-times for finding ∇u_h (if S.O.R. is applied for solving the linear system of equations) are .06, .041, 2.69, 20.1 sec. and for finding $\tilde{\nabla} u_h$ 0.1, 0.48, 2.9, 21 sec.

We finally note that formula (3.15) can be applied for smoothing the boundary flux $\Lambda = \frac{\partial}{\partial n} u|_{\partial\Omega}$. In above test example it gives the $O(h^2|\ln h|)$ accuracy for Λ. For further details we refer to [22].

4. THE NODAL INTERIOR $O(h^4)$-SUPERCONVERGENCE OF THE GRADIENT

In this section we combine results of [5] and [37] to obtain the $O(h^4)$-superconvergence of a postprocessed gradient at nodes of $\Omega_0 \subset\subset \Omega$ for the problem (4.1) below. Note that mostly only the $O(h^2)$-superconvergence of the gradient has been obtained employing the linear triangular elements (see Section 2).

Let $\{\mathcal{T}_h\}$ be a family of uniform triangulations consisting of right-angled isosceles triangles, $\bar{\Omega} = \bigcup_{T \in \mathcal{T}_h} T$, and let us deal with the problem:

$$\begin{cases} -\dfrac{\partial^2 u}{\partial x_1^2} + \dfrac{\partial^2 u}{\partial x_1 \partial x_2} - \dfrac{\partial^2 u}{\partial x_2^2} = f & \text{in } \Omega, \\ u = 0 & \text{on } \partial\Omega, \end{cases} \tag{4.1}$$

i.e. $A = \begin{pmatrix} 1 & -1/2 \\ -1/2 & 1 \end{pmatrix}$ and $a = 0$ in (3.1). Consider a linear one-to-one mapping (see Figure 4.1)

$$x = F(\hat{x}) = B\hat{x}, \qquad \hat{x} \in R^2,$$

where $B = \begin{pmatrix} \sqrt{3}/2 & -1/2 \\ 0 & 1 \end{pmatrix}$. Setting $\hat{f}(\hat{x}) = f(F(\hat{x}))$, $\hat{u}(\hat{x}) = u(F(\hat{x}))$, $\hat{\Omega} = F^{-1}(\Omega)$, we obtain that $\hat{u} \in H_0^1(\hat{\Omega})$ and $\nabla\hat{u}(\hat{x}) = B^T \nabla u(x)$. Thus from the weak formulation of (3.1) and the equality $A = B \cdot B^T$, we find that $\hat{u}$ is a weak solution of the problem

$$\begin{cases} -\Delta\hat{u} = \hat{f} & \text{in } \hat{\Omega}, \\ \hat{u} = 0 & \text{on } \partial\hat{\Omega}. \end{cases} \tag{4.2}$$

Defining a triangulation of $\hat{\Omega}$

$$\hat{T}_h = \{F^{-1}(T) \mid T \in T_h\} ,$$

which consists of equilateral triangles (see Figure 4.1), we can easily

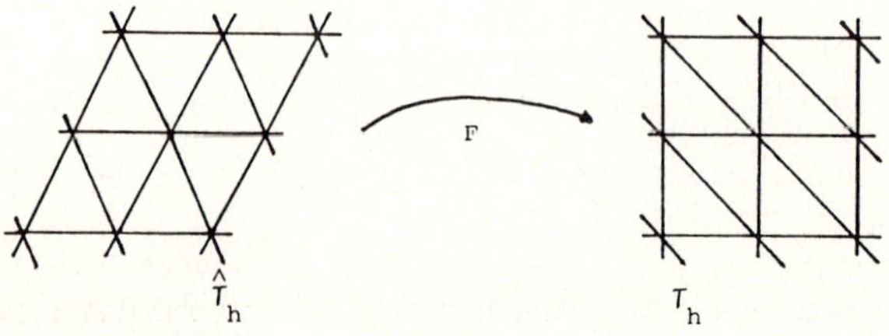

Figure 4.1.

ascertain that for the corresponding Galerkin approximation $\hat{u}_h$ of $\hat{u}$, it is $\hat{u}_h(\hat{x}) = u_h(x)$. Using further the definition relation $\hat{u}(\hat{x}) = u(x)$ and (2.5) (for $\hat{u}$ and $\hat{u}_h$), we get

$$|u - u_h|_{h,\Omega} \equiv \max_{x \in N_h \cap \Omega} |u(x) - u_h(x)| \le C h^4 \|u\|_{C^{4+\varepsilon}(\bar{\Omega})} , \tag{4.3}$$

where u is the solution of (4.1) with the reguired smoothness.

Note that an analogous result can be obtained for any constant symmetric and positive definite matrix A. Setting e.g. $B = \sqrt{A}$, we easily determine the shape of the triangles used.

Now we recall the result of [37], p. 61. For our case, it may be rewritten as follows

$$\max_{x \in N_h \cap \Omega_0} \|\nabla u(x) - \tilde{\nabla} u_h(x)\|_{R^2} \le C(h^k \|u\|_{7,\infty,\Omega_1} + |u - u_h|_{h,\Omega_1}) , \tag{4.4}$$

where $\Omega_0 \subset\subset \Omega_1 \subset\subset \Omega$ and k is the order of approximation of an operator $\tilde{\nabla}$. Defining $\tilde{\nabla} u_h$ for $x = (x_1, x_2) \in N_h \cap \Omega_0$ in this way

$$(\tilde{\nabla} u_h(x))_1 = \frac{1}{12h}(-u_h(x_1 + 2h, x_2) + 8u_h(x_1 + h, x_2) \tag{4.5}$$

$$- 8u_h(x_1 - h, x_2) + u_h(x_1 - 2h, x_2)) ,$$

(and the second component $(\tilde{\nabla} u_h(x))_2$ similarly), then clearly $k = 4$. Now, the combination of (4.3) and (4.4) yields the interior $O(h^4)$-superconvergence of the gradient at nodes.

Theorem 4.1. Let $\Omega \subset \mathbb{R}^2$ be bounded polygon which can be covered by uniform triangulation $\mathcal{T}$ consisting of right-angled isosceles triangles i.e.

$$\bar{\Omega} = \bigcup_{T \in \mathcal{T}} T .$$

Let $u \in C^{4+\varepsilon}(\bar{\Omega})$ $(\varepsilon > 0)$ be the weak solution of (4.1) such that $u|_{\Omega_1} \in W^7_\infty(\Omega_1)$ for some $\Omega_1 \subset\subset \Omega$. Then

$$\| \nabla u - \tilde{\nabla} u_h \|_{h,\Omega_0} \le C h^4 \left(\|u\|_{7,\infty,\Omega_1} + \|u\|_{C^{4+\varepsilon}(\bar{\Omega})} \right)$$

where $\Omega_0 \subset\subset \Omega_1$ and $\tilde{\nabla} u_h$ is the approximation of ∇u obtained from u_h by formula (4.5).

<u>Example 4.1.</u> Let $\Omega = (0,1) \times (0,1)$ and $\Omega_0 = (\frac{1}{4}, \frac{3}{4}) \times (\frac{1}{4}, \frac{3}{4})$, and let f in (4.1) be such

$$u(x_1, x_2) = x_2(x_2 - 1) \sin \pi x_1$$

is the exact solution. The same family of triangulations is used as in the test of Section 2 (see Figure 2.2). The results of the Table 4.1 confirm the theoretical rates of convergence.

<u>Table 4.1.</u>

h	$\lvert u - u_h \rvert_{h,\Omega}$	$\lvert (\nabla u - \tilde{\nabla} u_h)_1 \rvert_{h,\Omega_0}$	$\lvert (\nabla u - \tilde{\nabla} u_h)_2 \rvert_{h,\Omega_0}$
1/4	4.491494 E-4	-	-
1/8	2.728846 E-5	4.940120 E-4	5.163512 E-5
1/16	1.694235 E-6	3.122729 E-5	3.203487 E-6
1/32	1.057185 E-7	1.957230 E-6	2.003494 E-7
rate of convergence	$O(h^4)$	$O(h^4)$	$O(h^4)$

REFERENCES

[1] S. Agmon: Lectures on elliptic boundary value problems. Van Nostrand Mathematical Studies 2. D. Van Nostrand Company, Inc., New York, 1965.

[2] A.B. Andreev: Superconvergence of the gradient for linear triangle elements for elliptic and parabolic equations. C.R.Acad. Bulgare Sci. 37 (1984), 293-296.

[3] O. Axelsson and V.A. Baker: "Finite element solution of boundary value problems. Theory and computation", Academic Press, Computer Science and Applied Mathematics, Orlando, Florida, 1984.

[4] I. Babuška and A. Miller: The post-processing in the finite element method, Part I. Internat. J. Numer. Methods Engrg. 20 (1984), 1085-1109.

[5] H. Blum, Q. Lin and R. Rannacher: Asymptotic error expansion and Richardson extrapolation for linear finite elements. Numer. Math., to appear.

[6] F. Brezzi, J. Douglas and L.D. Marini: Two families of mixed finite elements for second order elliptic problems, Numer. Math. 47 (1985), 217-235.

[7] C.M. Chen: Optimal points of stresses of triangular linear element in the finite element method (Chinese). Natural Sci. J. Xiangtan Univ. 1 (1978), 77-90.

[8] C.M. Chen: Superconvergence of finite element approximations to nonlinear elliptic problems. (Proc. China-France Sympos. on Finite Element Methods, Beijing, 1982), Science Press, Beijing, Gordon and Breach Sci. Publishers, Inc., New York, 1983, 622-640.

[9] C.M. Chen: Superconvergence of finite element methods (Chinese). Advances in Math. 14 (1985), 39-51.

[10] M. Delfour, G. Payne and J.-P. Zolesio: An optimal triangulation for second order elliptic problems, Comput. Meth. Appl. Mech. Engrg., 50 (1985), 231-262.

[11] A.R. Diaz, N. Kikuchi and J.E. Taylor: A method of grid optimization for finite element methods. Comput. Meth. Appl. Mech. Engrg., 41 (1983), 29-45.

[12] Q.V. Dinh: Simulation numérique en éléments finis d'éqoulements de fluides visquex incompressibles par une méthode de décomposition de domaines sur processeurs vectoriels, Thése de 3e cycle, Université Pierre et Marie Curie, 1982.

[13] I. Fried: On the optimality of the pointwise accuracy of the finite element solution. Internat. J. Numer. Methods Engrg. 15 (1980), 451-456.

[14] W. Hackbusch: "Multigrid methods and applications", Springer series in Computational Mathematics, Vol. 4, Springer Verlag, Berlin, 1985.

[15] B. Heinrich: Integralbilanzmethode für elliptische Probleme, II. Fehlerabschätzung und Konvergenz. Beiträge Numer. Math. 12 (1984), 75-94.

[16] E. Hinton and J.S. Campbell: Local and global smoothing of discontinuous finite element functions using a least squares method. Internat. J. Numer. Methods Engrg. 8 (1974), 461-480.

[17] I. Hlaváček and M. Křížek: On a superconvergent finite element scheme for elliptic systems, I. Dirichlet boundary conditions, II. Boundary conditions of Newton's or Neumann's type, III. Optimal interior estimates, 1985, (submitted to Apl. Mat.).

[18] V.P. Iljin: Properties of some classes of differentiable functions of several variables in n-dimensional domain (Russian). Trudy Mat. Inst. Steklov. 66 (1962), 227-363.

[19] V.A. Kontratiev: Boundary problems for elliptic equations in domains with conical or angular points. Trans. Moscow Math. Soc. 16 (1967), 277-313.

[20] M. Křížek and P. Neittaanmäki: Superconvergence phenomenon in the finite element method arising from averaging gradients. Numer. Math. 45 (1984), 105-116.

[21] M. Křížek and P. Neittaanmäki: On superconvergence techniques. Preprint n. 34, Univ. of Jyväskylä, 1984, 1-43.

[22] M. Křížek and P. Neittaanmäki: On a global superconvergence of the gradient of linear triangular elements. J. Comp. Appl. Math., to appear.

[23] N. Levine: Pointwise logarithm-free error estimates for finite elements on linear triangles. Numer. Anal. Report n. 6/84, Univ. of Reading, 1984.

[24] N. Levine: Superconvergent recovery of the gradient from piecewise linear finite element approximations. Numer. Anal. Report n. 6/83, Univ. of Reading, 1983, 1-25.

[25] N. Levine: Superconvergent estimation of the gradient from linear finite element approximations on triangular elements. Numer. Anal. Report 3/85, Univ. of Reading, 1985, 1-202.

[26] Q. Lin: High accuracy from the linear elements. Proc. of the Fifth Beijing Sympos. on Differential Geometry and Differential Equations, 1984, 1-5.

[27] Q. Lin and J.Ch. Xu: Linear finite element with high accuracy, J. Comput. Math. 3, 1985, n. 1.

[28] Q. Lin and J. Wang: Some expansions of the finite element approximation. Research Report IMS-15. Chengdu Branch of Acad. Sinica, 1984, 1-11.

[29] Q. Lin and Q.D. Zhu: Asymptotic expansion for the derivative of finite elements. J. Comput. Math. 2 (1984), 361-363.

[30] J. Nečas: Les méthodes directes en théorie des équations elliptiques. Academia, Prague, 1967.

[31] P. Neittaanmäki and M. Křížek: Superconvergence of the finite element schemes arising from the use of averaged gradients. (Proc. Conf. ARFEC, Lisbon, 1984), Lisbon, 1984, 169-178.

[32] L.A. Oganesjan, V.J. Rivkind and L.A. Ruchovec: Variational-difference methods for the solution of elliptic equations (Russian), Part I. (Proc. Sem., Issue 5, Vilnius, 1973), Inst. of Phys. and Math., Vilnius, 1973, 3-389.

[33] R. Rannacher: On Richardson's extrapolation for FEM, These Proceedings.

[34] R. Rannacher and R. Scott: Some optimal error estimates for piecewise linear finite element approximations. Math. Comp. 38 (1982), 437-445.

[35] R. Stenberg: On the post-processing at equilibrium finite element methods, These proceedings.

[36] V. Thomée and B. Westergren: Elliptic difference equations and interior regularity. Numer. Math. 11 (1968), 196-210.

[37] B. Westergren: Interior estimates for elliptic systems of difference equations (thesis). Univ. of Göteborg, 1982, 1-100.

ON A SIMPLE FINITE ELEMENT METHOD FOR PLATE BENDING PROBLEMS

J. Pitkäranta
Institute of Mathematics
Helsinki University of Technology
SF-02150 Espoo, Finland

SUMMARY

We prove the convergence of a simple finite element method based on the Discrete Kirchhoff Triangle (DKT) for solving the Mindlin plate equations. If t is the thickness of the plate, an error bound $\mathcal{O}(h^2+t^2)$ is obtained for both the bending and the rotations.

INTRODUCTION

In the Mindlin plate model the total energy of the plate of thickness t is given by

$$F_t(u,\theta) = \int_\Omega [a(\theta,\theta) + \frac{1}{t^2} |\theta-\nabla u|^2 - 2fu]dx. \tag{1}$$

Here u and θ stand for the bending and rotations, respectively, f is proportional to the external (vertical) load, and a is a bilinear form satisfying

$$C|\nabla\theta|^2 \geq a(\theta,\theta) \geq c|\nabla\theta|^2$$

for some positive constants c and C. We assume here for simplicity that $a(\theta,\theta) = |\nabla\theta|^2$. Moreover we assume that Ω is a convex polygon and that the plate is clamped, i.e. $u = \theta = 0$ on $\partial\Omega$.

It is well known that a direct finite element technique of minimizing the energy F_t gives poor results if the thickness of the plate is much smaller than the mesh parameter h. As is also well understood, the failure is caused by the constraint $\nabla = \nabla u$, which is imposed asymptotically as $t \to 0$. In low-order finite element approximations this constraint causes the shear energy to be overestimated - the so called locking effect.

One of the simplest ways of avoiding locking is the shear correction trick of Fried [2]. The trick is simply to replace t by h (or a constant times h) in (1). The main purpose of this note is to show why the shear correction trick works. As a by-product we show also why a seemingly "cloudy" combination of shear correction with the so called "discrete Kirchhoff hypothesis" (another trick) can in fact lead to a reasonable method.

The analysis techniques to be presented can be extended to cover a rather wide range of plate elements, see [3]. Here we concentrate on the analysis of a simple low-order element, the Discrete Kirchhoff Triangle (DKT). In [1], DKT is named as one of the most effective low-order plate elements.

THE DISCRETE KIRCHHOFF TRIANGLE

In the Discrete Kirchhoff Triangle, the bending is approximated by a reduced cubic polynomial, with Hermitean degrees of freedom, while the rotations are linear.

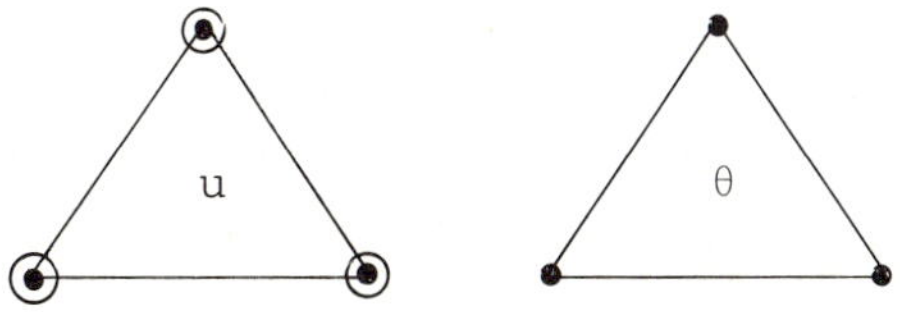

The Discrete Kirchhoff Triangle (DKT)

The basic steps in the implementation of the element are essentially those suggested in [2]:

Step 1. (Shear correction) Replace t by (a constant times) h in (1).

Step 2. (Discrete Kirchhoff hypothesis) Eliminate θ by imposing the constraint $\theta(x^i) = \nabla u(x^i)$ at the vertices x^i of the triangulation.

Let $V = H_0^1(\Omega)$, $W = V^2$, and let $V_h \subset V$ and $W_h \subset W$ be the finite element subspaces for the bending and the rotations, respectively, in the DKT-scheme. Then we are seeking a pair (u_h, θ_h) which minimizes F_h in the space

$$Z_h = \{(v,\varphi) \in V_h \times W_h : \varphi(x^i) = \nabla\varphi(x^i) \text{ at the vertices } x^i \text{ of the triangulation}\}.$$

Let (u^t, θ^t) be the exact solution of the plate bouding problem, i.e., the point in $V \times W$ where F_t is minimized. Let further $\|\cdot\|_s$, $s = -2,\ldots,3$ denote the norm of the Sobolev space $[H^s(\Omega)]^d$, $d = 1,2$, where for $s > 0$, $H^{-s}(\Omega)$ is the dual of $H_0^s(\Omega)$.

Theorem 1. For a given $f \in H^{-1}(\Omega)$ in (1), let (u^t, θ^t) and (u_h, θ_h) be defined as above for $t > 0$ and let $(u^0, \theta^0) = \lim_{t\to 0} (u^t, \theta^t)$. Then there is a constant C independent of t, h and f such that the following estimates hold:

$$\|u^t - u^0\|_1 + \|\theta^t - \theta^0\|_0 \le Ct^2 \|f\|_{-1}, \tag{2}$$

$$\|u^t - u^h\|_1 + \|\theta^t - \theta^h\|_0 \le C(t^2 + h^2)\|f\|_{-1}. \tag{3}$$

Estimate (2) states essentially that so far as the discretization error of the finite element scheme is not better than $O(h^2)$, alterning the parameter t in the range $0 \le t \le h$ does not substantially affect the overall approximation error. In particular, the seemingly contradictory assumptions made in implementing DKT ($t = h$ at Step 1, $t = 0$ at Step 2) are legitimate in this sense.

Let us prove estimate (2). To this end, we note first that if (u^t, θ^t) minimizes F_t, and if $\lambda^t = t^{-2}(\theta^t - \nabla u^t)$, then the triple $(u^t, \theta^t, \lambda^t)$ is the solution to

$$-\Delta\theta^t + \lambda^t = 0, \tag{4.a}$$

$$\operatorname{div} \lambda^t = f, \tag{4.b}$$

$$\theta^t - \nabla u^t = t^2 \lambda^t. \tag{4.c}$$

Here λ^t has the physical meaning of a shear stress. For $t = 0$ these equations are written equivalently as

$$\Delta^2 u^0 = f, \qquad \theta^0 = \nabla u^0, \qquad \lambda^0 = \Delta\theta^0,$$

and the boundary conditions then reduce to $u^0 = \frac{\partial u^0}{\partial n} = 0$ on $\partial\Omega$. This corresponds to the Kirchhoff plate model. We need below the regularity estimate

$$\|u^0\|_3 + \|\theta^0\|_2 + \|\lambda^0\|_0 \le C\|f\|_{-1}, \tag{5}$$

valid on a convex polygonal domain.

Eqs. (4) admit the weak formulation

$$(\nabla\theta^t, \nabla\varphi) + (\lambda^t, \varphi) = 0 \quad , \quad \varphi \in W, \tag{6.a}$$

$$-(\lambda^t, \nabla v) = (f, v) \quad , \quad v \in V, \tag{6.b}$$

$$(\theta^t - \nabla u^t, \mu) = t^2(\lambda^t, \mu), \quad \mu \in Q, \tag{6.c}$$

where $Q = [L_2(\Omega)]^2$ and $(\cdot,\cdot)$ stands for the inner product in $L_2(\Omega)$ or Q. It follows in particular that

$$(\nabla(\theta^t - \theta^0), \nabla\varphi) + (\lambda^t - \lambda^0, \varphi) = 0, \ \varphi \in W, \tag{7.a}$$

$$-(\lambda^t - \lambda^0, \nabla v) = 0, \ v \in V, \tag{7.b}$$

$$(\nabla(u^t - u^0) - (\theta^t - \theta^0), \mu) + t^2(\lambda^t - \lambda^0, \mu) = -t^2(\lambda^0, \mu), \mu \in Q. \tag{7.c}$$

Choosing here $\varphi = \theta^t - \theta^0$, $v = u^t - u^0$, $\mu = \lambda^t - \lambda^0$, and summing the equations we see that

$$\|\nabla(\theta^t - \theta^0)\|_0^2 + t^2\|\lambda^t - \lambda^0\|_0^2 = -t^2(\lambda^0, \lambda^t - \lambda^0),$$

and hence by (5)

$$\|\nabla(\theta^t - \theta^0)\|_0 \le Ct\|f\|_{-1}, \tag{8.a}$$

$$\|\lambda^t - \lambda^0\|_0 \le C\|f\|_{-1}. \tag{8.b}$$

We now improve estimate (8.a) using a duality argument. Let (ρ, ψ, ξ) be the solution to

$$(\nabla\psi,\nabla\varphi) + (\xi,\varphi) = (g_1,\varphi), \qquad \varphi \in W, \tag{9.a}$$

$$-(\xi,\nabla v) = (g_2,v), \qquad v \in V, \tag{9.b}$$

$$(\psi-\nabla\rho,\mu) = 0 \quad , \qquad \mu \in Q, \tag{9.c}$$

where $g_1 \in Q$ and $g_2 \in H^{-1}(\Omega)$. Then $\Delta^2\rho = -\operatorname{div} g_1 + g_2$, $\psi = \nabla\rho$, and $\xi = \Delta\psi + g_1$ on Ω and $\rho = \frac{\partial\rho}{\partial n} = 0$ on $\partial\Omega$, so we have the regularity estimate

$$\|\rho\|_3 + \|\psi\|_2 + \|\xi\|_0 \leq C(\|g_1\|_0 + \|g_2\|_{-1}). \tag{10}$$

Choose now $\varphi = \theta^t-\theta^0$, $v = u^t-u^0$, $\mu = \lambda^t-\lambda^0$ in Eqs. (9) and set $\varphi = -\psi$, $v = -\rho$, $\mu = -\xi$ in Eqs. (7). Then sum the resulting six equations to obtain the identity

$$0 = (g_1,\theta^t-\theta^0) + (g_2,u^t-u^0) + t^2(\lambda^0,\xi).$$

Since $g_1 \in Q$ and $g_2 \in H^{-1}(\Omega)$ are here arbitrary, the asserted estimate (2) follows recalling the regularity estimates (5) and (10).

Let us now prove estimate (3) for the DKT-scheme. Thus consider a finite element scheme where the minimum of F_h is sought in $Z_h \subset V_h \times W_h$, where V_h and W_h are finite element subspaces associated to a quasiuniform triangulation τ_h of Ω. Then if $Q_h \subset Q$ is another finite element subspace such that $Q_h \supset W_h \oplus \nabla V_h$ the scheme may be formulated equivalently as: Find $(u_h,\theta_h) \in Z_h$ and $\lambda_h \in Q_h$ such that

$$(\nabla\theta_h,\nabla\varphi) + (\lambda_h,\varphi-\nabla v) = (f,v), \quad (v,\varphi) \in Z_h, \tag{11.a}$$

$$(\theta_h-\nabla u_h,\mu) = h^2(\lambda_h,\mu), \qquad \mu \in Q_h. \tag{11.b}$$

Since we have in fact $\lambda_h = h^{-2}(\theta_h-\nabla u_h)$, it follows that

$$\|\nabla\theta_h\|_0^2 + h^2\|\lambda_h\|_0^2 = (f,u_h), \tag{12.a}$$

$$h^{-2}\|\theta_h-\nabla u_h\|_0^2 \leq (f,u_h). \tag{12.b}$$

In order to combine (12.a) and (12.b) into a stability estimate, we need some mesh-dependent norms. Let σ_h denote the collection of all sides in triangulation τ_h that are interior to Ω. Then if $\varphi \in [L_2(\Omega)]^2$ is such that $\varphi_{|T} \in [H^1(T) \cap C^0(\bar{T})]^2$ for all $T \in \tau_h$, we define

$$|\varphi|_{1,h}^2 = \sum_{T\in\tau_h} \int_T |\nabla\varphi|^2 dx + \sum_{S\in\sigma_h} h^{-1}|[\varphi]|^2 ds,$$

where $[\varphi]$ denotes the jump of φ across S. Further if $v \in H^1(\Omega)$ is such that $v_{|T} \in H^2(T) \cap C^1(\bar{T})$ for all $T \in \tau_h$, we define

$$\|v\|_{2,h}^2 = \|v\|_1^2 + |\nabla v|_{1,h}^2 .$$

Now, in view of (12a,b) we have

$$\|u_h\|_1^2 \le C(\|\theta_h-\nabla u_h\|_0^2 + \|\theta_h\|_0^2) \le C_1(f,u_h).$$

Similarly, applying a simple inverse estimate,

$$\begin{aligned}|\nabla u_h|_{1,h}^2 &\le 2|\theta_h-\nabla u_h|_{1,h}^2 + 2|\theta_h|_{1,h}^2 \\ &\le Ch^{-2}\|\theta_h-\nabla u_h\|_0^2 + 2\|\nabla\theta_h\|_0^2 \\ &\le C_1(f,u_h).\end{aligned}$$

Upon combining these estimates with (12a,b) we obtain the stability estimate

$$\|u_h\|_{2,h} + \|\theta_h\|_1 + h\|\lambda_h\|_0 \le C\|f\|_{-2,h}, \tag{13}$$

where

$$\|f\|_{-2,h} = \sup_{v\in V^h} \frac{(f,v)}{\|v\|_{2,h}}.$$

Note that in the continuous case one has for $t = 0$ the estimate

$$\|u^0\|_2 + \|\theta^0\|_1 + \|\lambda^0\|_{-1} \le C\|f\|_{-2},$$

so the norms appearing in (13) may be regarded as the natural norms for the plate problem.

Having established stability in the "right" norms, the error analysis is straightforward. Let

$$\mathcal{B}_t(u,\theta,\lambda;v,\varphi,\mu) = (\nabla\theta,\nabla\varphi) + (\lambda,\varphi-\nabla v) - (\theta-\nabla u,\mu) + t^2(\lambda,\mu).$$

Then

$$\mathcal{B}_h(u^0-u_h,\theta^0-\theta_h,\lambda^0-\lambda_h;v,\varphi,\mu) = -h^2(\lambda^0,\mu),\quad (v,\varphi,\mu)\in Z_h\times Q_h.$$

Upon combining this with the stability estimate (13), which can be stated equivalently as:

$$\sup_{(v,\varphi,\mu)\in Z_h\times Q_h} \frac{\mathcal{B}(u,\theta,\lambda;v,\varphi,\mu)}{\|(v,\varphi,\mu)\|_h} \ge C\|(u,\theta,\lambda)\|_h,\quad (u,\theta,\lambda)\in Z_h\times Q_h,$$

where $\|(v,\varphi,\mu)\|_h^2 = \|v\|_{2,h}^2 + \|\varphi\|_1^2 + h^2\|\mu\|_0^2$, it follows by standard reasoning that for the DKT-scheme the following estimate holds:

$$\|u^0-u_h\|_{2,h} + \|\theta^0-\theta_h\|_1 + h\|\lambda^0-\lambda_h\| \le Ch\|f\|_{-1}.$$

Moreover, a duality argument gives

$$\|u^0-u_h\|_1 + \|\theta^0-\theta_h\|_0 \le Ch^2\|f\|_{-1}.$$

Together with (2), this proves the asserted error bound (3).

Let us finally point out that estimate (3) holds more generally if for any $u \in H^3(\Omega) \cap V$, $\theta \in [H^2(\Omega)]^2 \cap W$ there are interpolants $\tilde{u} \in V_h$ and $\tilde{\theta} \in W_h$ satisfying:

a) $\|u-\tilde{u}\|_{2,h} \leq Ch\|u\|_3$,

$\|\theta-\tilde{\theta}\|_1 \leq Ch\|\theta\|_2$,

b) $\theta = \nabla u \Rightarrow (\tilde{u},\tilde{\theta}) \in Z_h$.

Here requirement a) means that while linear elements are sufficient for rotations, one should take at least quadratic elements for the bending.

REFERENCES

[1] BATOZ, J.-L., BATHE, K.-J., LEE-WING-HO: "A study of three-node triangular plate bending elements", Int.J.Numer.Meth. Engrg. 15 (1980), pp. 1771-1812.

[2] FRIED, I., SHOK KENG YANG: "Triangular, nine-degrees-of-freedom, C^0 plate bending element of quadratic accuracy", Quart. Appl. Math. 31 (1973), pp. 303-312.

[3] PITKÄRANTA, J.: "Analysis of finite element methods for solving the Mindlin plate equations", to appear.

RICHARDSON EXTRAPOLATION WITH FINITE ELEMENTS

R. Rannacher

FB Mathematik, Universität des Saarlandes

D-66oo Saarbrücken 11, W. Germany

SUMMARY

In a recent paper by H. Blum, Lin Q., and the author,[1], it has been shown that the Ritz projection method with linear finite elements admits an asymptotic error expansion for certain classes of "uniform" meshes. This provides the theoretical justification for the use of Richardson extrapolation or related correction processes for increasing the accuracy of the scheme. The proofs are entirely based on finite element techniques and extend to situations where no "discrete" maximum principle is available. This is examplarily demonstrated here for the second-order Lamé-Navier system in plain linear elasticity and for a mixed formulation of the fourth-order Kirchhoff plate bending model.

I. INTRODUCTION

Let us first consider the usual model problem

$$- \Delta u = f \quad \text{in } \Omega \ , \quad u = g \quad \text{on } \partial\Omega \ , \tag{1.1}$$

on a bounded domain $\Omega \subset R^2$. It is well known that the standard 5-point finite difference discretization of (1.1) combined with a higher order boundary approximation admits an asymptotic error expansion, [18], [3], [12],

$$u_h(z) = u(z) + h^2 e_1(z) + h^4 e_2(z) + \dots \ , \tag{1.2}$$

with functions $e_k(z)$ independent of the mesh size parameter h. The expansion (1.2) is the theoretical justification of Richardson extrapolation or related correction processes for increasing the accuracy of the scheme,[12]. The derivation of such error expansions for finite difference schemes is usually based on the "discrete" maximum principle. In fact, the range of the validity of (1.2) has been extended as far as the classical "discrete" maximum principle reaches, [4]. It is the purpose of this paper to describe how error expansions may be derived by using finite element techniques for cases where no "discrete" maximum principle is available:

a) strongly coupled elliptic systems as that of plain linear elasticity;

b) fourth-order problems as those occuring in linear plate bending theory.

In [8], [9], [1o], and [1] it has been shown that the Ritz projection method with linear finite elements applied to the model problem (1.1) admits an asymptotic error expansion if the triangulation is uniform in a certain sense. The results of [1] can be summarized as follows:

Let Ω be a convex polygonal domain and let the triangulation be "three-directional", i.e., the sides of all triangles are parallel to three fixed direction unit vectors. Then, if the solution u is sufficiently smooth, there holds

$$u_h(z) = u(z) + h^2 e_1(z) + h^4 e_2(z) + o(h^4) , \qquad (1.3)$$

in nodal points z uniformly bounded away from the corner points of $\partial\Omega$. If the triangulation is only "piecewise" three-directional with respect to some macro-triangulation of Ω, then

$$u_h(z) = u(z) + h^2 e_1(z) + O(h^4 \ln \frac{1}{h}) \qquad (1.4)$$

holds in nodal points z uniformly bounded away from the corner points of $\partial\Omega$ and from the vertices of the macro-triangulation. In the case of a curved boundary $\partial\Omega \in C^2$, the expansion (1.4) remains valid with a remainder term of the order $O(h^3 \ln 1/h)$ in interior nodal points if the triangulation is uniform in the interior of Ω and properly refined along the boundary.

The proofs are based on the "discrete" Green representation formula

$$(u_h - u)(z) = \sum_P g_h(P) \int_\Omega \nabla(u - i_h u) \cdot \nabla N_P \, dx , \qquad (1.5)$$

where g_h is the discrete analogue of the Green function g of problem (1.1), $i_h u$ is the piecewise linear interpolant of u, and N_P are the usual nodal basis functions of the finite element space. The quantities under the sum on the right hand side of (1.5) represent the consistency error in the scheme and are expanded with respect to powers of h. The Green function g_h is controlled via sharp estimates for L^1-error terms of the form $\|g - g_h\|_1 = O(h^2 \ln^2 1/h)$ which are partly available from [6].

Notation. Below, $L^p(\Omega)$, $W^{m,p}(\Omega)$, $H^m(\Omega)$, and $\overset{o}{H}{}^m(\Omega)$ are the usual Lebesgue- and Sobolev spaces, and $\|\cdot\|_{m,p} = \|\cdot\|_{m,p;\Omega}$, $\|\cdot\|_p = \|\cdot\|_{o,p;\Omega}$, etc., the corresponding norms; $(\cdot,\cdot)$ and $\|\cdot\|$ are the inner product and norm of $L^2(\Omega)$ or $L^2(\Omega)^2$. For later purposes, we fix any two subdomains $\Omega' \subset \Omega'' \subset \Omega$ such that $\mathrm{dist}(\Omega', \Delta_\Omega) > \mathrm{dist}(\Omega'', \Delta_\Omega) > o$, where Δ_Ω is the set of all corner points of $\partial\Omega$. The partial derivatives are written as $\partial_\nu = \frac{\partial}{\partial x_\nu}$, $\nu = 1,2$.

II. SOME EXPANSION FORMULAS

Let $\Omega \subset R^2$ be a convex polygonal domain, and let $\mathcal{T}_h = \{T\}$ be a uniform triangulation of Ω of width h, which is generated by three direction unit vectors. For any fixed triangle $T \in \mathcal{T}_h$, we introduce the notation

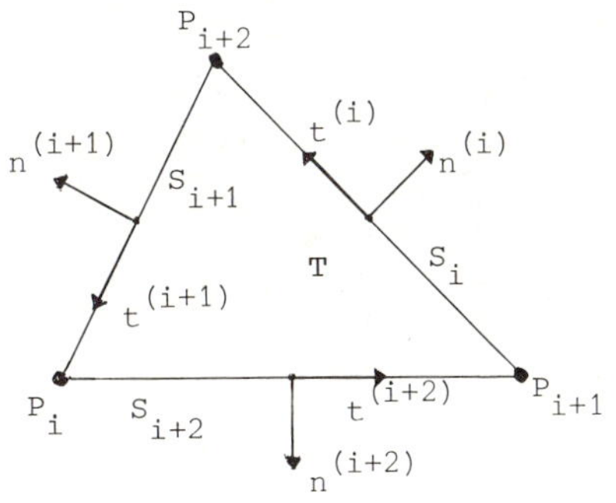

P_i vertices of T ("nodal points")

S_i side of T opposite to P_i

$h_i = \lambda_i h$ length of S_i

$A = \alpha h^2$ area of T

$n^{(i)}$ outer normal unit vector along S_i

$t^{(i)}$ tangent unit vector along S_i

$D_i = t^{(i)} \cdot \nabla$ directional derivative .

Here and below, the indices $i, i+1, i+2$ are to be understood modulo 3 .

Corresponding to $\mathcal{T}_h$ we consider the finite element spaces

$$S_h = \{\phi_h \in H^1(\Omega) \ , \ \phi_h \text{ linear on each } T \in \mathcal{T}_h\} \ , \quad S_{h,o} = S_h \cap H^1_o(\Omega) \ .$$

We shall derive asymptotic expansion formulas for integrals of the type

$$\int_\Omega \partial_\nu \phi_h \partial_\mu (v - i_h v) \, dx \quad , \quad \int_\Omega \phi_h (v - i_h v) \, dx \quad , \tag{2.1}$$

where $\phi_h \in S_h$, v smooth, and $i_h v \in S_h$ such that $i_h v(P) = v(P)$ for all nodal points P of $\mathcal{T}_h$.

For any fixed triangle $T \in \mathcal{T}_h$ there holds

$$\int_T \partial_\nu \phi_h \partial_\mu (v - i_h v) \, dx = \sum_{i=1}^3 n_\mu^{(i)} \int_{S_i} \partial_\nu \phi_h (v - i_h v) \, ds \equiv \sum_{i=1}^3 n_\mu^{(i)} I_\nu^{(i)} \ . \tag{2.2}$$

The line integrals $I_\nu^{(i)}$ will be expanded in powers of h following the approach in [1]. Let $N_i \in S_h$ denote the element nodal basis function corresponding to the vertex P_i, $N_i(P_j) = \delta_{ij}$. On T, there holds

$$\sum_{j=1}^3 \partial_\nu N_j = o \quad , \quad \partial_\nu N_i = - \frac{h_i}{2A} n_\nu^{(i)} \quad ,$$

and, consequently, for $i \in \{1,2,3\}$,

$$\partial_\nu \phi_h = \sum_{j=1}^3 \phi_h(P_j) \partial_\nu N_j = \{\phi_h(P_{i+2}) - \phi_h(P_{i+1})\} \partial_\nu N_{i+2} + \{\phi_h(P_i) - \phi_h(P_{i+1})\} \partial_\nu N_i$$

$$= \frac{h_i h_{i+1}}{2A} \{ n_\nu^{(i)} D_{i+2} \phi_h - n_\nu^{(i+2)} D_i \phi_h \} \ .$$

Combining this with the identity

$$\int_{S_i} (v - i_h v) \, ds = - \frac{h_i^2}{12} \int_{S_i} D_i^2 v \, ds + h_i^2 \int_{S_i} \sigma_i D_i^4 v \, ds \quad , \tag{2.3}$$

where $\sigma_i(s) = s^2 h_i^{-2} (1 - s h_i^{-1})^2$, leads us to

$$I_\nu^{(i)} = \frac{h_i h_{i+2}^3}{24A} \{ n_\nu^{(i+2)} D_i \phi_h \int_{S_i} D_i^2 v \, ds - n_\nu^{(i)} D_{i+2} \phi_h \int_{S_i} D_i^2 v \, ds + \tag{2.4}$$
$$+ h_i^4 \int_{S_i} \sigma_i \partial_\nu \phi_h D_i^4 v \, ds \ .$$

By the theorem of Gauss and the relations

$$t^{(i+1)} \cdot n^{(i)} = -t^{(i)} \cdot n^{(i+1)} = - \frac{2A}{h_i h_{i+1}} \quad ,$$

one sees that, for any $w \in W^{1,1}(T)$,

$$h_i \int_{S_{i+2}} w \, ds - h_{i+2} \int_{S_i} w \, ds = \frac{h_1 h_2 h_3}{2A} \int_T D_{i+1} w \, dx \quad .$$

We use this in (2.4) to replace the derivative $D_{i+2}\phi_h$ on S_i by $D_i \phi_h$. Inserting the result into (2.2) and rearranging terms eventually yields

$$\int_T \partial_\nu \phi_h \partial_\mu (v - i_h v) \, dx = h^2 \frac{\lambda_1 \lambda_2 \lambda_3}{48 \alpha^2} \int_T \phi_h \{ D_1 D_2 D_3 \sum_{i=1}^3 n_\nu^{(i)} n_\mu^{(i)} \lambda_i^3 D_i \} v \, dx +$$
$$+ h^4 \sum_{i=1}^3 \lambda_i^4 n_\mu^{(i)} \int_{S_i} \sigma_i \partial_\nu \phi_h D_i^4 v \, ds + \ldots \tag{2.5}$$

$$\dots + \frac{h^2}{24\alpha}\sum_{i=1}^{3}\left\{\int_{S_i} D_i\phi_h[\lambda_i^3\lambda_{i+2}n_\mu^{(i)}n_\nu^{(i+2)}D_i^2-\lambda_{i+1}^4 n_\mu^{(i+1)}n_\nu^{(i+1)}D_{i+1}^2]v\,ds\,+\right.$$

$$\left. + \int_{S_i}\phi_h[\lambda_i^3\lambda_{i+1}n_\mu^{(i)}n_\nu^{(i)}D_{i+1}D_i-\lambda_{i+2}^4 n_\mu^{(i+2)}n_\nu^{(i+2)}D_iD_{i+2}^2]v\,ds\right\} .$$

Next, we take any $\hat{T} \in \mathcal{T}_h$ to figure as a reference triangle for the whole triangulation and denote by Γ_i (i=1,2,3) that part of $\partial\Omega$ which is parallel to the side S_i of $\hat{T}$. With this notational convention we sum (2.5) for all $T \in \mathcal{T}_h$ and, by the cancellation of several line integrals, obtain the following expansion formula

$$\int_\Omega \partial_\nu\phi_h\partial_\mu(v-i_hv)\,dx = h^2\varepsilon_\Omega^{\nu\mu}(\phi_h,v) + h^2\varepsilon_\Gamma^{\nu\mu}(\phi_h,v) + h^4\rho_h^{\nu\mu}(\phi_h,v) , \tag{2.6}$$

where

$$\varepsilon_\Omega^{\nu\mu}(\phi_h,v) = \frac{\lambda_1\lambda_2\lambda_3}{48\alpha^2}\int_\Omega \phi_h[D_1D_2D_3\sum_{i=1}^{3}n_\mu^{(i)}n_\nu^{(i)}\lambda_i^3D_i]v\,dx$$

$$\varepsilon_\Gamma^{\nu\mu}(\phi_h,v) = \frac{1}{24\alpha}\sum_{i=1}^{3}\int_{\Gamma_i}\{D_i\phi_h[\lambda_i^3\lambda_{i+2}n_\mu^{(i)}n_\nu^{(i+2)}D_i^2-\lambda_{i+1}^4 n_\mu^{(i+1)}n_\nu^{(i+1)}D_{i+1}^2]v\,+$$

$$+\,\phi_h[\lambda_i^3\lambda_{i+1}n_\mu^{(i)}n_\nu^{(i)}D_{i+1}D_i-\lambda_{i+2}n_\mu^{(i+2)}n_\nu^{(i+2)}D_iD_{i+2}^2]v\}ds$$

$$\rho_h^{\nu\mu}(\phi_h,v) = \sum_{T\in\mathcal{T}_h}\sum_{i=1}^{3}\lambda_i^4 n_\mu^{(i)}\int_{S_i}\sigma_i\partial_\nu\phi_h D_i^4 v\,ds .$$

We note that, for $\phi_h \in S_{h,o}$ and $v \in C^4(\bar{\Omega})$,

$$\varepsilon_\Gamma^{\nu\mu}(\phi_h,v) = o . \tag{2.7}$$

Furthermore, by integration by parts on Γ_i, one sees that, for $\phi_h \in S_h$ and $v \in C^4(\bar{\Omega}) \cap H_o^2(\Omega)$,

$$\varepsilon_\Gamma^{\nu\mu}(\phi_h,v) = \frac{1}{24\alpha}\sum_{i=1}^{3}\int_{\Gamma_i}\phi_h[\lambda_{i+1}^4 n_\mu^{(i+1)}n_\nu^{(i+1)}D_{i+1}^2 - \dots \dots -\lambda_{i+2}^4 n_\mu^{(i+2)}n_\nu^{(i+2)}D_{i+2}^2]D_iv\,ds ; \tag{2.8}$$

notice that the functions in square brackets vanish in the corner points of the boundary $\partial\Omega$.

<u>Lemma 2.1.</u> <u>Let</u> $v \in C^4(\bar{\Omega})$ <u>and</u> $\phi_h \in S_h$, <u>and suppose that there exists some function</u> $\phi \in W^{2,1}(\Omega)$, <u>such that</u>

$$h^{-1}\|\phi-\phi_h\|_{1,1} + \|\phi\|_{2,1} \leq \tau_h . \tag{2.9}$$

<u>Then, there holds, uniformly for</u> $h > o$,

$$\int_\Omega \partial_\nu\phi_h\partial_\mu(v-i_hv)\,dx = h^2\{\varepsilon_\Omega^{\nu\mu}(\phi_h,v) + \varepsilon_\Gamma^{\nu\mu}(\phi_h,v)\} + O(h^4\tau_h) . \tag{2.10}$$

<u>Proof.</u> For the remainder term in (2.6) we have

$$|\rho_h^{\nu\mu}(\phi_h,v)| = |\rho_h^{\nu\mu}(\phi_h-\phi,v) + \rho_h^{\nu\mu}(\phi,v)|$$

$$\leq c\left\{\sum_{T\in\mathcal{T}_h}\int_{\partial T}|\nabla(\phi-\phi_h)|ds + \int_{\partial\Omega}|\nabla\phi|ds\right\}\|v\|_{4,\infty} ,$$

and consequently, by a standard argument,

$$|\rho_h^{\nu\mu}(\phi_h,v)| \le c\,\{h^{-1}\,\|\phi-\phi_h\|_{1,1} + \|\phi\|_{2,1}\}\,\|v\|_{4,\infty} . \qquad \square$$

In order to derive an expansion formula for the second integral expression in (2.1), we need an error representation for the 2-dimensional trapezoidal rule over triangles. There exists an extensive literature on the extension of the classical Euler-MacLaurin formula to n-dimensional versions of the trapezoidal rule; see [11]. However, we prefer to give an elementary direct proof for the following expansion result:

Lemma 2.2. For $T \in \mathcal{T}_h$ and $w \in W^{4,1}(T)$, there holds

$$\int_T w\,dx = \frac{\alpha h^2}{3}\sum_{i=1}^{3} w(P_i) - \frac{h^2}{24}\int_T \sum_{i=1}^{3}\lambda_i^2 D_i^2 w\,dx + \tag{2.11}$$

$$+\frac{h^4}{180}\,\frac{\alpha}{\lambda_1\lambda_2\lambda_3}\int_{\partial T}\sum_{i=1}^{3}\lambda_i^2 D_i^2(\lambda_{i-1}D_{i-1}-\lambda_{i+1}D_{i+1})w\,dx + h^4\,r_T(w) ,$$

where $|r_T(w)| \le c\,\|w\|_{4,1;T}$.

Proof. Let $\hat{T}$ be the unit triangle with the vertices $P_1=(0,0)$, $P_2=(1,0)$, $P_3=(0,1)$, and q any cubic polynomial. By a straightforward use of the one-dimensional formula (2.3), one derives the identity

$$\int_0^1\int_0^{1-x_2} q\,dx_1dx_2 = \frac{1}{6}\{q(0,0)+q(1,0)+q(0,1)\} - \frac{1}{12}\int_0^1\int_0^{1-x_2}\{\partial_1^2 q-\partial_1\partial_2 q+\partial_2^2 q\}dx_1dx_2$$

$$+\frac{1}{360}\int_0^1\{2\partial_1^3 q-3\partial_1^2\partial_2 q-3\partial_1\partial_2^2 q+2\partial_2^3 q\}\,dy .$$

Observing that $|\hat{T}| = 1/2$, $|\partial\hat{T}| = 2+\sqrt{2}$, $h_1=\sqrt{2}$, $h_2=h_3=1$, and $D_1=(\partial_2-\partial_1)/\sqrt{2}$, $D_2=-\partial_2$, $D_3=\partial_1$, we obtain the representation

$$\int_{\hat{T}} q\,dx = \frac{|\hat{T}|}{3}\sum_{i=1}^{3} q(P_i) - \frac{1}{24}\int_{\hat{T}}\sum_{i=1}^{3} h_i^2 D_i^2 q\,dx +$$

$$+\frac{1}{180}\,\frac{|\hat{T}|}{|\partial\hat{T}|}\int_{\partial\hat{T}}\sum_{i=1}^{3} h_i^2 D_i^2(h_{i-1}D_{i-1}-h_{i+1}D_{i+1})q\,dx ,$$

which is invariant under affin-linear transformations. Hence, it holds for any triangle $T \in \mathcal{T}_h$. For general $w \in W^{4,1}(T)$, the assertion then follows by a standard argument. $\square$

Lemma 2.3. Let $v \in W^{4,1}(\Omega)$ and $\phi_h \in S_h$, and suppose that there exists some $\phi \in W^{2,1}(\Omega)$, such that (2.9) holds. Then, there holds, uniformly in h,

$$\int_\Omega \phi_h(v-i_h v)\,dx = h^2\,\varepsilon_\Omega^{00}(\phi_h,v) + O(h^4\tau_h) , \tag{2.12}$$

where

$$\varepsilon_\Omega^{00}(\phi_h,v) = -\frac{1}{24}\int_\Omega \phi_h\sum_{i=1}^{3}\lambda_i^2 D_i^2 v\,dx .$$

Proof. Using Lemma 2.2 with $w = \phi_h(v-i_h v)$, we have

$$\int_\Omega \phi_h(v-i_h v)\,dx = -\frac{h^2}{24}\sum_{T\in\mathcal{T}_h}\int_T \sum_{i=1}^{3}\lambda_i^2 D_i^2[\phi_h(v-i_h v)]\,dx +$$

$$+\frac{h^4}{180}\frac{\alpha}{\lambda_1\lambda_2\lambda_3}\sum_{T\in\mathcal{T}_h}\int_{\partial T}\sum_{i=1}^{3}\lambda_i^2 D_i^2(\lambda_{i-1}D_{i-1}-\lambda_{i+1}D_{i+1})[\phi_h(v-i_h v)]\,ds +$$

$$+h^4\sum_{T\in\mathcal{T}_h} r_T(\phi_h[v-i_h v]) = \Sigma_1+\Sigma_2+\Sigma_3 .$$

For the first sum, Σ_1, there holds

$$\Sigma_1 = -\frac{h^2}{24}\int_\Omega \phi_h\sum_{i=1}^{3}\lambda_i^2 D_i^2 v\,dx - \frac{h^2}{12}\int_\Omega\sum_{i=1}^{3}\lambda_i^2 D_i(\phi_h-\phi)D_i(v-i_h v)\,dx -$$

$$-\frac{h^2}{12}\sum_{i=1}^{3}\lambda_i\left\{\int_{\partial\Omega} D_i(\phi_h-\phi)t^{(i)}\cdot n(v-i_h v)\,ds - \int_\Omega D_i^2\phi(v-i_h v)\,dx\right\},$$

and, consequently, by well-known estimates for $v-i_h v$,

$$\Sigma_1 = h^2\,\varepsilon_\Omega^{oo}(\phi_h,v) + O(h^4\tau_h) .$$

The second and third sum can be estimated as follows

$$|\Sigma_2| \le c\,h^4\sum_{T\in\mathcal{T}_h}\int_{\partial T}\{|\nabla\phi_h||\nabla^2 v| + |\phi_h||\nabla^3 v|\}\,ds = O(h^4\tau_h) ,$$

$$|\Sigma_3| \le c\,h^4\sum_{T\in\mathcal{T}_h}\|\phi_h(v-i_h v)\|_{4,1;T} = O(h^4\tau_h) .$$

This clearly proves the representation (2.12). □

III. ERROR EXPANSION FOR AN ELLIPTIC SYSTEM

As a model situation we consider the Lamé-Navier system of plain linear elasticity,

$$(\lambda+\mu)\,\mathrm{grad\,div}\,\underline{u} + \mu\Delta\underline{u} = \underline{f} \quad \text{in } \Omega , \quad \underline{u} = \underline{b} \quad \text{on } \partial\Omega , \tag{3.1}$$

where $\Omega \subset R^2$ is assumed to be convex polygonal. Here, $\underline{u} = (u^1,u^2)$ is the unknown displacement, $\underline{b} = (b^1,b^2)$ and $\underline{f} = (f^1,f^2)$ are the prescribed boundary displacement and the body force, respectively, both assumed to be sufficiently regular; λ and μ are the Lamé constants. We use the abbreviation $V = H^1(\Omega)^2$ and $V_o = H_o^1(\Omega)^2$ for the spaces of admissible vector functions. The strongly coupled system (3.1) is V_o-elliptic, i.e., the corresponding energy form

$$a(\underline{u},\underline{\phi}) = \int_\Omega\{\lambda\,\mathrm{tr}(\varepsilon[\underline{u}])\mathrm{tr}(\varepsilon[\underline{\phi}]) + 2\mu\varepsilon[\underline{u}]:\varepsilon[\underline{\phi}]\}\,dx ,$$

where $\varepsilon[\underline{v}] = 1/2(\nabla\underline{v} + \nabla\underline{v}^T)$, satisfies a Korn inequality on V_o. Hence, the variational formulation of (3.1),

$$\underline{u} \in V_o + \underline{b} \quad : \qquad a(\underline{u},\underline{\phi}) = (\underline{f},\underline{\phi}) \quad , \quad \forall\, \underline{\phi} \in V_o , \tag{3.2}$$

has a unique solution $\underline{u}$ which is in $W^{2,p}(\Omega)^2$, for some $p > 2$, and smooth in the subdomain $\Omega'' \subset \Omega$. For notational convenience, we shall write the energy form in the more general form

$$a(\underline{u},\underline{\phi}) = \sum_{j,k=1}^{2} \sum_{\nu,\mu=1}^{2} (a^{jk}_{\nu\mu}\partial_\nu u^j, \partial_\mu \phi^k) .$$

For discretizing problem (3.1) we consider the usual displacement method with linear shape functions. Let $\mathcal{T}_h = \{T\}$ be a regular triangulation of Ω of width h, and

$$V_h = \{ \underline{\phi}_h \in V ,\ \phi^\alpha_h \in S_h \ (\alpha=1,2) \} \ , \quad V_{h,o} = V_h \cap V_o .$$

The discrete analogue of (3.2) is

$$\underline{u}_h \in V_{h,o} + i_h \underline{b} \quad : \quad a(\underline{u}_h,\underline{\phi}_h) = (\underline{f},\underline{\phi}_h) \quad , \quad \forall\, \underline{\phi}_h \in V_{h,o} . \tag{3.3}$$

For this approximation scheme we have the standard L^2-error estimate

$$\|\underline{u}-\underline{u}_h\| + h\, \|\underline{u}-\underline{u}_h\|_{1,2} \le c\, h^2 \|\underline{u}\|_{2,2} , \tag{3.4}$$

and the "interior" L^∞-error estimate, [14],

$$\|\underline{u}-\underline{u}_h\|_{\infty;\Omega'} \le c\, h^2 \ln\frac{1}{h} \{ \|\underline{u}\|_{2,\infty;\Omega''} + \|\underline{u}\|_{2,2} \} . \tag{3.5}$$

If the triangulation $\mathcal{T}_h$ is uniform, the estimate (3.5) can be extended to an asymptotic error expansion. To this end, let $z \in \Omega'$ be any fixed nodal point, and let $G^z_h = [\underline{g}^1_h,\underline{g}^2_h] \in V_{h,o} \times V_{h,o}$ be the "discrete" Green matrix of problem (3.3) which is defined through the relation

$$a(\underline{\phi}_h,\underline{g}^\alpha_h) = \phi^\alpha_h(z) \quad , \quad \forall\, \underline{\phi}_h \in V_{h,o} , \quad \alpha \in \{1,2\} . \tag{3.6}$$

Observing that $\underline{u}_h - i_h\underline{u} \in V_{h,o}$, we obtain the discrete Green representation formula

$$(u^\alpha_h - u^\alpha)(z) = a(\underline{u}_h - i_h\underline{u},\underline{g}^\alpha_h) = a(\underline{u} - i_h\underline{u},\underline{g}^\alpha_h) . \tag{3.7}$$

<u>Theorem 3.1.</u> <u>Suppose that $\underline{u} \in C^{4+\varepsilon}(\bar{\Omega})$, for some $\varepsilon > o$, and that the triangulation $\mathcal{T}_h$ is three-directional. Then, at nodal points $z \in \Omega'$, there holds</u>

$$\underline{u}_h(z) = \underline{u}(z) + h^2\, \underline{e}(z) + O(h^4 \ln\frac{1}{h}) , \tag{3.8}$$

<u>where the function $\underline{e}(z)$ is independent of h.</u>

<u>Proof.</u> Let $G^z = [\underline{g}^1,\underline{g}^2]$ denote the "continuous" Green matrix of problem (3.1) corresponding to the point $z \in \Omega'$. We want to apply Lemma 2.1 with $\phi_h = g^{\alpha j}_h$. To this end, we introduce a "regularized" Green matrix $\tilde{G}^z \in V_o \times V_o$ which is defined through the relation

$$a(\underline{\phi},\underline{\tilde{g}}^\alpha) = (\phi^\alpha,\tilde{\delta}) \quad , \quad \forall\, \underline{\phi} \in V_o ,$$

where $\tilde{\delta}$ is a smooth approximation of the Dirac functional at z with

$\operatorname{diam}(\operatorname{supp}\tilde{\delta}) = O(h)$; see [6] and [1] for an analogous construction in the scalar case. For this one can prove the estimate

$$h^{-1}\|g_h^\alpha - \tilde{g}^\alpha\|_{1,1} + \|\tilde{g}^\alpha\|_{2,1} = O(\ln\frac{1}{h}) . \tag{3.9}$$

For homogeneous material, $a_{\nu\mu}^{jk} \equiv$ const., there holds

$$a(\underline{u}-i_h\underline{u},\underline{g}_h^\alpha) = \sum_{j,k=1}^{2}\sum_{\nu,\mu=1}^{2} a_{\nu\mu}^{jk}\int_\Omega \partial_\nu g_h^{\alpha j}\partial_\mu(u^\alpha - i_h u^\alpha)\,dx .$$

Hence, by Lemma 2.1 we obtain that, for $\underline{u} \in C^4(\bar{\Omega})$,

$$a(\underline{u}-i_h\underline{u},\underline{g}_h^\alpha) = h^2\,\varepsilon(\underline{g}_h^\alpha,\underline{u}) + O(h^4\ln\frac{1}{h}) , \tag{3.1o}$$

where, in view of (2.7),

$$\varepsilon(\underline{g}_h,\underline{u}) = \sum_{j,k=1}^{2}\sum_{\nu,\mu=1}^{2} a_{\nu\mu}^{jk}\,\varepsilon_\Omega^{\nu\mu}(g_h^{\alpha j},u^k) .$$

The functions

$$e_h^\alpha(z) = \varepsilon(\underline{g}_h^\alpha,\underline{u}) , \quad e^\alpha(z) = \varepsilon(\underline{g}^\alpha,\underline{u}) \quad (\alpha = 1,2)$$

are the components of the solutions of the problems (3.2) and (3.3), respectively, corresponding to the force term $(\underline{f},\underline{\phi}) = \varepsilon(\underline{\phi},\underline{u})$ and the boundary values $\underline{b} = o$. Hence, by the "interior" error estimate (3.5), there holds

$$(\underline{e}-\underline{e}_h)(z) = O(h^2\ln\frac{1}{h}) . \tag{3.11}$$

Combining (3.11), (3.1o), and (3.7), we obtain that, for $\underline{u} \in C^{4+\varepsilon}(\bar{\Omega})^2$,

$$(u_h^\alpha - u^\alpha)(z) = h^2 e^\alpha(z) + O(h^4\ln\frac{1}{h}) \qquad (\alpha = 1,2) . \qquad \square$$

Remarks. Analogously as in the scalar case, [1], expansions of the form (3.8) can also be proven if the triangulation $\mathcal{T}_h$ is only piecewise uniform or, in the case of a curved boundary, if $\mathcal{T}_h$ is only uniform up to a boundary strip of width h . Further, by a refined analysis of the remainder term in (3.1o), the expansion (3.8) can be shown to remain valid if the solution $\underline{u}$ is in $W^{4,p}(\Omega)^2 \cap C^{4+\varepsilon}(\bar{\Omega}'')^2$, for some $p > 2$ and $\varepsilon > o$. If $\underline{u}$ has only the minimal regularity of $W^{2,p}(\Omega)^2 \cap C^{2+\varepsilon}(\bar{\Omega}'')^2$, then (3.8) still remains valid with a remainder term of the order $o(h^2)$.

IV. ERROR EXPANSION FOR A FOURTH-ORDER PROBLEM

We consider the model boundary value problem

$$\Delta^2 u = f \quad \text{in} \quad \Omega , \quad u = \partial_n u = o \quad \text{on} \quad \partial\Omega . \tag{4.1}$$

Here, **u** may describe the deflection of a thin elastic plate subjected to the load f ; $\Omega \subset \mathbb{R}^2$ is the middle surface of the plate which is clamped at the boundary $\partial\Omega$. Again, Ω is assumed to be convex polygonal and f to be sufficiently smooth. For discretizing problem (4.1), we take a simple *mixed* finite element method which treats **u** and $v = -\Delta u$ simultaneously as unknowns. There exists a unique weak solution $u \in H_o^2(\Omega)$ of (4.1) which is in $W^{4,q}(\Omega) \cap C^{6+\varepsilon}(\bar{\Omega}'')$, for some $q > 1$ depending on the maximum interior angle of $\partial\Omega$. Clearly, the pair $\{u,v\} \in H_o^1(\Omega) \times H^1(\Omega)$ is the (unique) solution of the *mixed* variational system

$$(\nabla\psi,\nabla u) - (\psi,v) + (\nabla v,\nabla\phi) = F(\phi) + B(\psi) , \qquad (4.2)$$

for all $\{\phi,\psi\} \in H^1_o(\Omega) \times H^1(\Omega)$, where here $F(\phi) = (f,\phi)$ and $B(\psi) = o$.

Let again $\mathcal{T}_h = \{T\}$ be a regular triangulation of Ω of width h , and $S_h \subset H^1(\Omega)$, $S_{h,o} \subset H^1_o(\Omega)$ the corresponding spaces of piecewise linear finite elements. The discrete analogue of (4.2),

$$(\nabla\psi_h,\nabla u_h) - (\psi_h,v_h) + (\nabla v_h,\nabla\phi_h) = F(\phi_h) + B(\psi_h) , \qquad (4.3)$$

for all $\{\phi_h,\psi_h\} \in V_{h,o} \times V_h$, uniquely determines approximations $\{u_h,v_h\}$ in $V_{h,o} \times V_h$ to $\{u,v\}$. Several variants of this approximation scheme have been analyzed first in [7] and [5]. For the present case of *linear* shape functions, the convergence estimate

$$\|u-u_h\| + h^{1/2}\|v-v_h\| \leqq c\,h^{1-\varepsilon}\|u\|_{4,2} \qquad (4.4)$$

has been shown in [13], for (interior) uniform meshes, and later on in [16], for general meshes, together with the pointwise estimate

$$\|u-u_h\|_\infty \leqq c\,h^{1-\varepsilon}\|u\|_{4,2} . \qquad (4.5)$$

Furthermore, from [17] one has the "interior" result

$$\|v-v_h\|_{\Omega'} \leqq c\{h^{2-\varepsilon}\|u\|_{4,2;\Omega''} + h^{-\varepsilon}\|u-u_h\|_2\} . \qquad (4.6)$$

Below, we shall use the following refinements of the estimates (4.4)-(4.5) which hold on three-directional triangulations

$$\|u-u_h\| + h\|v-v_h\| \leqq c\,h^2\{\|u\|_{3,2} + \|f\|\} , \qquad (4.7)$$

$$\|u-u_h\|_{\infty;\Omega'} \leqq c\,h^{2-\varepsilon}\{\|u\|_{3,\infty;\Omega''} + \|u\|_{4,q}\} ; \qquad (4.8)$$

the proofs of these results will be given in a forthcoming paper, [15].

Substracting (4.3) from (4.2) leads to the error identity

$$(\nabla\psi_h,\nabla[u-u_h]) - (\psi_h,u-u_h) + (\nabla[v-v_h],\phi_h) = o , \qquad (4.9)$$

for all $\{\phi_h,\psi_h\} \in V_{h,o} \times V_h$. In order to represent the error $(u-u_h)(z)$ at some nodal point z , we use again "discrete" Green functions $\{g_h,\theta_h\}$ in $V_{h,o} \times V_h$ which are defined through the equation

$$(\nabla\psi_h,\nabla g_h) - (\psi_h,\theta_h) + (\nabla\theta_h,\nabla\phi_h) = \phi_h(z) , \qquad (4.1o)$$

for all $\{\phi_h,\psi_h\} \in V_{h,o} \times V_h$. Taking $\phi_h = u_h - i_h u$ in (4.1o), we obtain the error representation formula

$$(u_h-u)(z) = (\nabla[v-i_h v],\nabla g_h) - (v-i_h v,\theta_h) + (\nabla\theta_h,\nabla[u-i_h u]) . \qquad (4.11)$$

__Theorem 4.1. Suppose that__ $u \in C^6(\bar{\Omega})$, __and that the triangulation__ $\mathcal{T}_h$ __is three-directional. Then, at nodal points__ $z \in \Omega'$, __there holds__

$$u_h(z) = u(z) + h^2 e(z) + O(h^{3.5}) , \qquad (4.12)$$

__where the function__ $e(z)$ __is independent of__ h .

Proof. Let $g = g^z \in H_o^2(\Omega) \cap W^{3,q}(\Omega)$, $1 \leq q < 2$, be the "continuous" Green function of problem (4.1) corresponding to the nodal point $z \in \Omega'$, and let $\theta = -\Delta g$. Furthermore, let again $\tilde{g} \in H_o^2(\Omega) \cap W^{4,1}(\Omega)$ be a regularized Green function defined through the equation

$$(\Delta\phi, \Delta\tilde{g}) = (\phi, \tilde{\delta}) \quad , \quad \forall \ \phi \in H_o^2(\Omega) \ ,$$

where $\tilde{\delta}$ is a smooth approximation of the Dirac functional; compare with the proof of Theorem 3.1 . There holds

$$h^{-1} \|g_h - \tilde{g}\|_{1,2} + \|\tilde{g}\|_{3,2} = O(\ell n \frac{1}{h}) \ , \tag{4.13}$$

and, for $\theta = -\Delta g$,

$$h^{-1} \|\theta_h - \tilde{\theta}\|_{1,1} + \|\tilde{\theta}\|_{2,1} = O(h^{-1/2}) \ ; \tag{4.14}$$

the proofs of these estimates will be contained in [15]. On the basis of (4.13) and (4.14) we can apply Lemmas 2.1 and 2.3 to obtain the relations

$$(\nabla[v - i_h v], \nabla g_h) = h^2 \varepsilon_\Omega(g_h, v) + O(h^4 \ell n \frac{1}{h}) \ , \tag{4.15}$$

$$(\nabla\theta_h, \nabla[u - i_h u]) = h^2 \{\varepsilon_\Omega(\theta_h, u) + \varepsilon_\Gamma(\theta_h, u)\} + O(h^{3.5}) \ , \tag{4.16}$$

where $\varepsilon_\Omega = \varepsilon_\Omega^{11} + \varepsilon_\Omega^{22}$, $\varepsilon_\Gamma = \varepsilon_\Gamma^{11} + \varepsilon_\Gamma^{22}$, and further,

$$(v - i_h v, \theta_h) = h^2 \varepsilon_\Omega^{oo}(\theta_h, v) + O(h^{3.5}) \ . \tag{4.17}$$

The functions

$$e(z) = \varepsilon_\Omega(g, v) - \varepsilon_\Omega^{oo}(\theta, v) + \varepsilon_\Omega(\theta, u) + \varepsilon_\Gamma(\theta, u) \ ,$$
$$e_h(z) = \varepsilon_\Omega(g_h, v) - \varepsilon_\Omega^{oo}(\theta_h, v) + \varepsilon_\Omega(\theta_h, u) + \varepsilon_\Gamma(\theta_h, u)$$

are the displacement components of the solutions of equations (4.2) and (4.3), respectively, corresponding to the inhomogeneous terms

$$F(\phi) = \varepsilon_\Omega(\phi, v) \quad , \quad B(\psi) = -\varepsilon_\Omega^{oo}(\psi, v) + \varepsilon_\Omega(\psi, u) + \varepsilon_\Gamma(\psi, u) \ .$$

By a careful analysis of the regularity of the functionals $F(\cdot)$ and $B(\cdot)$, one sees that $e \in W^{4,q}(\Omega)$, for some $q > 1$. Hence, by a slight generalization of the L^∞-estimate (4.8), to cover the case $B(\cdot) \not\equiv o$, it follows that

$$(e - e_h)(z) = O(h^{2-\varepsilon}) \ . \tag{4.18}$$

Combining (4.18) with (4.15)-(4.17) proves (4.12). □

Remarks. Again, expansions of the form (4.12) can also be proven if the triangulation $\mathcal{T}_h$ is only piecewise uniform, or, in the case of a curved boundary, if $\mathcal{T}_h$ is only uniform in the interior up to a boundary strip of width $O(h)$. Furthermore, the regularity requirements on u can be substantially weakened. The "minimum" regularity for which the expansion (4.12) holds with a remainder term of the order $o(h^2)$ should be $u \in W^{4,q}(\Omega)$, for some $q > 1$. The validity of expansions for the error $v - v_h$ is still an open question since for this quantity we do not even have optimal second-order convergence estimates in L^2.

Numerical example (from [2])

A clamped square plate with side-length 1 is subjected to the load f, resulting in the deflection $u(x,y) = 256x^2(1-x)^2y^2(1-y)^2$. Two mixed methods have been used on this test problem:

i) the Ciarlet/Raviart scheme, [5], as discussed in this paper;

ii) the Herrmann/Miyoshi scheme, [13], also with continuous linear shape functions, which yields approximations to u and to the single second derivatives $v^{jk} = \partial_j\partial_k u$ (j,k=1,2).

Let u_h denote the discrete solution corresponding to the mesh size h. From u_h, we obtain improved approximations $\bar{u}_h$ by h^2-extrapolation,

$$\bar{u}_h = \frac{1}{3}\{4u_{h/2} - u_h\} .$$

For the L^2-error and for the error at the mid-point c of the unit square, we observed the following orders, m, of convergence:

i) Ciarlet/Raviart scheme

h	$\|u-u_h\|$	$(u-u_h)(c)$	$(u-\bar{u}_h)(c)$	$\|v-v_h\|$	$(v-v_h)(c)$	$(v-\bar{v}_h)(c)$
1/4	.2418	.4786	.001354	2.802	4.573	.019541
1/8	.0608	.1186	.000216	.789	1.129	.001223
1/16	.0153	.0298	.000023	.201	.283	.000180
1/32	.0039	.0075	.000002	.050	.071	.000013
1/64	.0010	.0017	—	.012	.017	—
m	2	2	> 3.5	2	2	> 3.5

ii) Herrmann/Miyoshi scheme

h	$\|u-u_h\|$	$(u-u_h)(c)$	$(u-\bar{u}_h)(c)$	$\|v^{12}-v_h^{12}\|$	$(v^{12}-v_h^{12})(c)$	$(v^{12}-\bar{v}_h^{12})(c)$
1/4	.2532	.53377	.01927	7.924	-3.593	-.32522
1/8	.0698	.14789	.00346	2.963	- .654	-.02532
1/16	.0188	.03957	.00053	1.082	- .145	-.00128
1/32	.0049	.01029	.00007	.388	- .035	-.00007
1/64	.0013	.00262	—	.138	- .009	—
m	2	2	≧ 3	≧ 1.5	2	~ 4

These results indicate that the estimate for the remainder term in the expansion (4.12) may not be of optimal order, at least on a rectangular region. Further, the derivative quantities v seem to admit asymptotic error expansions, too; this effect will be studied in more detail in [2] and [15] .

REFERENCES

[1] BLUM,H., LIN,Q., RANNACHER,R.: Asymptotic error expansion and Richardson extrapolation for linear finite elements; to appear in Numer.Math.

[2] BLUM,H., RANNACHER,R.: A note on Herrmann's second mixed plate element; preprint 1986.

[3] BÖHMER,K.: Asymptotic expansions for the discretization error in linear elliptic boundary value problems on general regions; Math.Z. 177,235-255(1981).

[4] BÖHMER,K.: High order difference methods for quasi-linear elliptic boundary value problems on general regions; MRC Report, University of Wisconsin, Madison, 1979.

[5] CIARLET,P.G., RAVIART,P.A.: A mixed finite element method for the biharmonic equation; Mathematical Aspects of Finite Elements in Partial Differential Equations (C.de Boor,ed.),pp. 125-145, Academic Press, New York, 1974.

[6] FREHSE,J., RANNACHER,R.: Eine L^1-Fehlerabschätzung für diskrete Grundlösungen in der Methode der finiten Elemente; Finite Elemente,pp. 92-114, Bonn.Math.Schr. 89 (1976).

[7] GLOWINSKI,R.: Approximations externes, par éléments finis de Lagrange d'ordre un et deux, du problème de Dirichlet pour l'operateur biharmonique. Methodes itératives de résolution des problèmes approchés; Topics in Numerical Analysis(J.J.H.Miller,ed.),pp. 123-171, Academic Press, London, 1973.

[8] LIN,Q., LU,T.: Asymptotic expansions for finite element approximation of elliptic problems on polygonal domains; Sixth Int.Conf.Comp.Math. Appl.Sci.Eng., Versailles, 1983.

[9] LIN,Q., WANG,J.P.: Some expansions of the finite element approximation; Research Report IMS-15, Chengdu Branch of Academia Sinica, 1984.

[1o] LIN,Q, ZHU,Q.: Asymptotic expansion for the derivative of finite elements; J.Comput.Math. 2,361-363(1984).

[11] LYNESS,J.N., GENZ,A.C.: On simplex trapezoidal rule families; SIAM J. Numer.Anal. 17,126-147(198o).

[12] MARCHUK,G., SHAIDUROV,V.: Difference Methods and their Extrapolations; Springer, Berlin, 1983.

[13] MIYOSHI,T.: A finite element method for the solution of 4-th order partial differential equations; Kumamoto J.Sc.Math. 9,87-116(1973).

[14] NITSCHE,J.A.: Finite element approximations for solving the elasticity problem; Computing Methods in Applied Sciences and Engineering(R.Glowinski and J.L.Lions,ed.),pp. 154-166, Springer, Heidelberg, 1976.

[15] RANNACHER,R.: Estimates for discrete biharmonic Green functions and an asymptotic error analysis for Herrmann's second mixed plate element; preprint 1986.

[16] SCHOLZ,R.: A mixed method for 4th order problems using linear finite elements; R.A.I.R.O.Anal.Numér. 12, 85-9o(1978).

[17] SCHOLZ,R.: Interior error estimates for a mixed finite element method; Numer.Funct.Anal.Optim. 1,415-429(1979).

[18] WASOW,W.: Discrete approximations to elliptic differential equations; Z.Angew.Math.Phys. 6, 81-97(1955).

ON THE POSTPROCESSING OF MIXED EQUILIBRIUM FINITE ELEMENT METHODS

R. Stenberg

Swedish School of Economics and Business Administration

Arkadiagatan 22, SF-00100 Helsinki, Finland

SUMMARY

We consider mixed equilibrium finite element methods for second order elliptic boundary value problems. A new postprocessing method for improving the original variables is introduced and analyzed.

INTRODUCTION

In this note we will consider mixed finite element approximations of the so called equilibrium type. For simplicity of exposition we will consider the linear elasticity equations for a homogeneous isotropic body in a state of plane strain clamped at the boundary. The results are, however, trivially true in three dimensions, for other types of boundary conditions and also for mixed approximations of some other elliptic problems such as e.g. the Poisson problem. In the mixed approximation of the elasticity problem independent approximations are used for the stress tensor and the displacement.

In a recent paper Arnold and Brezzi [1] showed that the mixed method can be implemented by introducing Lagrange multipliers to enforce the interelement continuity of the normal component of the stress tensor. The main reason for introducing these artificial variables was to obtain an efficient way of solving the discretized equations. They showed that local condensation techniques could be used to eliminate the original variables, leading to a positive definite system for the Lagrange multipliers alone, which is a clear advantage when solving the system. However, they also showed that these Lagrange multipliers have the interpretation as approximations to the displacement and that they can be utilized in simple postprocessing procedures to construct a new displacement field which is considerably more accurate than the original one.

The main disadvantage of the approach sketched above is the relatively high number of parameters needed, even for low order approximations. An approach, to overcome this problem, advocated by us in an earlier paper [8] is to impose higher continuity requirements on the stress tensor and solving the discretized equations by penalty techniques. In this approach the postprocessing of [1] cannot, however, be used since the artificial Lagrange multipliers are not available. In this paper we will introduce a new postprocessing method which does not depend on the existence of a Lagrange multiplier approximation to the displacement along the interelement boundaries. Hence it can be used in connection with traditional mixed methods (cf. e.g. [1], [2], [4], [6]) as well as with those proposed in [8].

Let us also mention that there is another approach, suggested by us in [9], which can be used to achieve the same goal as that of the postprocessing techniques of [1] and this paper. For the details we refer directly to [9].

ANALYSIS

The model problem under consideration can be stated as: Given the body force $f = (f_1, f_2)$ find the symmetric stress tensor $\sigma = \{\sigma_{ij}\}$, and displacement $u = (u_1, u_2)$ such that

$$\begin{aligned} \sigma &= 2\mu\varepsilon(u) + \lambda\, \mathrm{tr}[\varepsilon(u)]\,\delta && \text{in } \Omega, \\ \operatorname{div}\sigma + f &= 0 && \text{in } \Omega, \\ u &= 0 && \text{on } \Gamma, \end{aligned} \tag{1}$$

where $\varepsilon(u) = \{\varepsilon_{ij}\}$, $i,j = 1,2$,

$$\varepsilon_{ij} = \frac{1}{2}\left(\frac{\partial u_i}{\partial x_j} + \frac{\partial u_j}{\partial x_i}\right),$$

is the deformation tensor,

$$\mathrm{tr}[\varepsilon(u)] = \varepsilon_{11} + \varepsilon_{22},$$

$$\operatorname{div}\sigma = \left(\frac{\partial\sigma_{11}}{\partial x_1} + \frac{\partial\sigma_{12}}{\partial x_2},\ \frac{\partial\sigma_{12}}{\partial x_1} + \frac{\partial\sigma_{22}}{\partial x_2}\right)$$

and δ is the unit tensor. Ω is assumed to be a plane polygonal domain with boundary Γ. $\mu > 0$ and $\lambda \geq 0$ are the Lamé constants.

The mixed method is based on the direct variational formulation of (1): Find $(\sigma, u) \in H \times V$ such that

$$\begin{aligned} a(\sigma,\tau) + (\operatorname{div}\tau, u) &= 0 && \tau \in H, \\ (\operatorname{div}\sigma, v) + (f, v) &= 0 && v \in V, \end{aligned} \tag{2}$$

where

$$V = [L^2(\Omega)]^2,$$

$$H = \{\sigma \in [L^2(\Omega)]^4 \mid \sigma_{ij} = \sigma_{ji},\ i,j = 1,2,\ \operatorname{div}\sigma \in V\}$$

$$a(\sigma,\tau) = \int_\Omega \left[\frac{1}{2\mu}\,\sigma\cdot\tau - \frac{\lambda}{4\mu(\mu+\lambda)}\,\mathrm{tr}(\sigma)\,\mathrm{tr}(\tau)\right]dx,$$

$$\sigma\cdot\tau = \sum_{i,j=1}^{2} \sigma_{ij}\tau_{ij}$$

and $(\cdot,\cdot)$ denotes the inner product in $[L^2(\Omega)]^2$.

The mixed method is now stated as: Find $(\sigma_h, u_h) \in H_h \times V_h \subset H \times V$ such that

$$\begin{aligned} a(\sigma_h, \tau) + (\operatorname{div} \tau, u_h) &= 0 \qquad && \tau \in H_h, \\ (\operatorname{div} \sigma_h, v) + (f,v) &= 0 && v \in V_h. \end{aligned} \tag{3}$$

H_h and V_h are typical finite element spaces based on a regular partitioning C_h of $\bar{\Omega}$ into triangles or quadrilaterals. As is usual (cf. [1], [2], [4], [8]) we will assume that

$$[P_k(K)]^4 \cap H_{|K} \subset H_{h|K} \qquad K \in C_h,$$

and

$$V_{h|K} = \begin{cases} [P_{k-1}(K)]^2 & \text{for } k \geq 2, \\ R(K) & \text{for } k = 1, \quad K \in C_h, \end{cases}$$

where $P_\ell(K)$ is the space of polynomials of degree ℓ on K and R(K) is the space of rigid body motions on K. Further, we will assume that the *equilibrium condition* is valid i.e.:

If $\tau \in H_h$ and $(\operatorname{div} \tau, v) = 0$ for every $v \in V_h$, then $\operatorname{div} \tau = 0$ in Ω. (4)

In order to carry out some duality arguments for the analysis of the postprocessing method we have to assume the regularity estimate

$$\| u \|_2 + \| \sigma \|_1 \leq C \| f \|_0 \tag{5}$$

for the solution (σ, u) to (1). This estimate is valid if the singularities due to the corners of Ω (or the change of boundary condition) are not too severe (cf. [7], [10] for the form of these singularities). Above and in the sequel we use standard Sobolev space notation (cf. [5]).

In our earlier papers [8], [9] we showed that the mixed method (3) can easily be analysed using the spaces

$$\mathcal{H}_h = \{\sigma = \{\sigma_{ij}\},\ i,j = 1,2.\ |\ \sigma_{ij} = \sigma_{ji},\ i,j = 1,2,\ \sigma \in [L^2(\Omega)]^4 \quad \sigma \cdot n \in [L^2(T)]^2 \quad T \in \Gamma_h\},$$

$$\mathcal{V}_h = \{u = (u_1, u_2)\ |\ u \in [L^2(\Omega)]^2,\ \varepsilon(u)_{|K} \in [L^2(K)]^4 \quad K \in C_h\}$$

with norms

$$\| \sigma \|_{0,h}^2 = \| \sigma \|_0^2 + \sum_{\substack{T \in \Gamma_h \\ T \subset \Gamma}} h_T \int_T |\sigma \cdot n|^2 \, ds$$

and

$$\| u\|^2_{1,h} = \sum_{K\in C_h} \| \varepsilon(u)\|^2_{0,K} + \sum_{T\in\Gamma_h} h_T^{-1} \int_T (u^+-u^-)^2\, ds$$

$$+ \sum_{T\subset\Gamma} h_T^{-1} \int_T u^2\, ds \quad ,$$

respectively. Here T denotes an edge of an element of C_h and h_T stands for the length of T. Γ_h denotes the collection of edges in the interior of Ω and

$$u^{\pm}(x) = \lim_{\varepsilon\to 0^{\pm}} u(x + \varepsilon n) \qquad x \in T \in \Gamma_h,$$

where n is the normal to T.

The interpolation estimates in the norms $\|\cdot\|_{0,h}$ and $\|\cdot\|_{1,h}$ are the same as those for the L^2-norm and the first order Sobolev norm, respectively.

Let us also note that an integration by parts on each $K \in C_h$ yields

$$(\mathrm{div}\ \sigma,v) \leq \|\sigma\|_{0,h} \|v\|_{1,h} \qquad \sigma \in H_h,\ v \in V_h.$$

The error estimates for the mixed method can now be collected as follows:

Theorem 1. Suppose that the method (3) satisfies the stability inequality

$$\sup_{0\neq\tau\in H_h} \frac{(\mathrm{div}\ \tau,v)}{\|\tau\|_{0,h}} \geq C\| v\|_{1,h} \qquad v \in V_h.$$

Then we have the error estimate

$$\| \Pi u - u_h\|_{1,h} + \| \sigma - \sigma_h\|_{0,h} \leq Ch^{k+1} |\sigma|_{k+1}. \tag{6}$$

If the regularity estimate (5) is valid we also have

$$\| u - u_h\|_0 \leq Ch^k \| u\|_k \tag{7}$$

and

$$\| \Pi u - u_h\|_0 \leq \begin{cases} Ch^2 \| u\|_2 & \text{for } k = 1, \\ Ch^{k+2} \| u\|_{k+2} & \text{for } k \geq 2, \end{cases} \tag{8}$$

where Π is the L^2-projection from V onto V_h.

Remark 1. Usually the stability of the method is stated using the norms of H and V (cf. [1], [2], [4]). It is, however, usually easy to show that those methods are also stable in the above norms.

Remark 2. The L^2-estimates for stress tensor contained in (6) and the estimates (7) and (8) can be obtained by an analysis based on the norms of H and V (cf. e.g. [4]). For our purpose we, however, need the estimate for the normal component of the stress tensor included in (6) and derived in [8], [9]. Likewise we will need the estimate (8). For completeness we will show how the mesh dependent norms can be utilized for the

Proof of (8). Let $(\gamma,z) \in H\times V$ be the solution to

$$a(\gamma,\tau) + (\operatorname{div}\tau,z) = 0 \qquad \tau\in H,$$
$$(\operatorname{div}\gamma,v) + (\Pi u - u_h,v) = 0 \qquad v\in V. \tag{9}$$

Choosing $\tau = \sigma - \sigma_h$ and $v = \Pi u - u_h$ we obtain in the usual manner

$$\|\Pi u - u_h\|_0^2 = (\tilde{\gamma} - \gamma,\sigma - \sigma_h) + (\operatorname{div}(\sigma - \sigma_h),\Pi z - z)$$
$$+ (\operatorname{div}(\tilde{\gamma} - \gamma),\Pi u - u_h) + (\operatorname{div}\tilde{\gamma},u - \Pi u), \tag{10}$$

where $\tilde{\gamma}\in H_h$ is the interpolant to γ. Due to the equilibrium condition the last term in the right hand side of (10) vanishes and we get

$$\|\Pi u - u_h\|_0^2 \le \|\gamma - \tilde{\gamma}\|_0 \, \|\sigma - \sigma_h\|_0 +$$
$$\|\sigma - \sigma_h\|_{0,h} \, \|z - \Pi z\|_{1,h} + \|\gamma - \tilde{\gamma}\|_{0,h} \, \|\Pi u - u_h\|_{1,h}$$
$$\le C(\|\gamma - \tilde{\gamma}\|_{0,h} + \|z - \Pi z\|_{1,h}) \, (\|\sigma - \sigma_h\|_0 +$$
$$\|\Pi u - u_h\|_{1,h}). \tag{11}$$

We now have the interpolation estimate

$$\|\gamma - \tilde{\gamma}\|_{0,h} + \|z - \Pi z\|_{1,h} \le \begin{cases} C(\|\gamma\|_0 + \|z\|_1) & \text{for } k=1, \\ Ch(\|\gamma\|_1 + \|z\|_2) & \text{for } k\ge 2. \end{cases} \tag{12}$$

The desired estimate now follows from (11), (12) and the regularity estimate

$$\|z\|_2 + \|\gamma\|_1 \le C \, \|\Pi u - u_h\|_0 \tag{13}$$

obtained from (5).

///

We are now ready to introduce our *postprocessing procedure*.

Let

$$V_h^* = \{v\in[L^2(\Omega)]^2 \mid v_{|K} \in [P_{k+1}(K)]^2 \; K\in C_h\}.$$

The new approximation $u_h^* \in V_h^*$ to u is now determined separately on each $K\in C_h$ by solving the problem:

$$(u_h^*,v)_K = (u_h,v)_K \qquad v\in V_{h|K}, \tag{14a}$$

$$b_K(u_h^*, v) = \int_{\partial K} \sigma_h \cdot n \, v \, ds + \int_K f \, v \, dx$$

$$v \in [V_h^* \cap V_h^\perp]|_K \quad , \qquad (14b)$$

where

$$b_K(u,v) = \int_K [\mu \nabla u \nabla v + (\lambda + \mu) \operatorname{div} u \operatorname{div} v] dx$$

and σ_h is the approximate stress tensor.

Remark 3. This means that u_h^* is calculated by first taking $\Pi u_h^* = u_h$ and then solving $(I - \Pi) u_h^*$ from

$$b_K((I - \Pi) u_h^*, v) = -b_K(u_h, v) + \int_{\partial K} \sigma_h \cdot n \, v \, ds$$

$$+ \int_K f \, v \, dx \qquad v \in [V_h^* \cap V_h^\perp]|_K$$

on each $K \in C_h$.

Since $R(K) \subset V_h|_K$ Korn's inequality implies the existence and uniqueness of u_h^*. We now prove that u_h^* indeed approximates u with a higher order of accuracy than u_h.

Theorem 2. Suppose that the regularity inequality (5) is valid. We then have

$$\| u - u_h^* \|_0 \leq \begin{cases} Ch^2 \| u \|_2 & \text{for } k = 1, \\ Ch^{k+2} \| u \|_{k+2} & \text{for } k \geq 2. \end{cases}$$

Proof. Let $\tilde{u} \in V_h^*$ be the L^2-projection of u. Since $\Pi(\tilde{u} - u_h^*) = \Pi u - u_h$ (8) gives

$$\| \Pi(\tilde{u} - u_h^*) \|_0 \leq \begin{cases} Ch^2 \| u \|_2 & \text{for } k = 1, \\ Ch^{k+2} \| u \|_{k+2} & \text{for } k \geq 2. \end{cases} \qquad (15)$$

Next, Korn's inequality implies

$$|(I - \Pi)(\tilde{u} - u_h^*)|_{1,K}^2 \leq b_K((I - \Pi)(\tilde{u} - u_h^*), (I - \Pi)(\tilde{u} - u_h^*))$$

$$= b_K(\tilde{u} - u_h^*, (I - \Pi)(\tilde{u} - u_h^*)) - b_K(\Pi(\tilde{u} - u_h^*), (I - \Pi)(\tilde{u} - u_h^*)).$$

Since the displacement u satisfies the variational eqation

$$b_K(u,v) = \int_{\partial K} \sigma \cdot n \, v \, ds + \int_K f \, v \, dx \qquad v \in [H^1(K)]^2,$$

we further get

$$|(I - \Pi)(\tilde{u} - u_h^*)|_{1,K}^2 \leq b_K(\tilde{u} - u, (I - \Pi)(\tilde{u} - u_h^*)) +$$

$$+ \int_{\partial K}(\sigma - \sigma_h)\cdot n((I - \Pi)(\tilde{u} - u_h^*))\, ds - b_K(\Pi(\tilde{u} - u_h^*), (I - \Pi)(\tilde{u} - u_h^*))$$

$$\leq C\{|u - \tilde{u}|_{1,K}|(I - \Pi)(\tilde{u} - u_h^*)|_{1,K} +$$

$$(h_K \int_{\partial K} |(\sigma - \sigma_h)\cdot n|^2\, ds)^{1/2} \cdot (h_K^{-1} \int_{\partial K} [(I - \Pi)(\tilde{u} - u_h^*)]^2\, ds)^{1/2}$$

$$+ |\Pi(\tilde{u} - u_h^*)|_{1,K}|(I - \Pi)(\tilde{u} - u_h^*)|_{1,K}\}. \qquad (16)$$

By normal scaling arguments one gets

$$(h_K^{-1} \int_{\partial K} [(I - \Pi)(\tilde{u} - u_h^*)]^2\, ds)^{1/2} \leq C|(I - \Pi)(\tilde{u} - u_h^*)|_{1,K},$$

$$|\Pi(\tilde{u} - u_h^*)|_{1,K} \leq Ch_K^{-1} \| \Pi(\tilde{u} - u_h^*)\|_{0,K}$$

and

$$h_K^{-1} \| (I - \Pi)(\tilde{u} - u_h^*)\|_{0,K} \leq C|(I - \Pi)(\tilde{u} - u_h^*)|_{1,K}.$$

Using these three inequalities (16) yields

$$\| (I - \Pi)(\tilde{u} - u_h^*)\|_{0,K}^2 \leq C\{h_K^3 \int_{\partial K} |(\sigma - \sigma_h)\cdot n|^2\, ds +$$

$$\| \Pi(\tilde{u} - u_h^*)\|_{0,K}^2\}.$$

Summing over all $K \in C_h$ we arrive at

$$\| (I - \Pi)(\tilde{u} - u_h^*)\|_0 \leq C\{h\| \sigma - \sigma_h\|_{0,h} + \| \Pi(\tilde{u} - u_h^*)\|_0\}$$

which together with (15) and (6) proves the asserted estimate.

///

REFERENCES

[1] Arnold, D. N., Brezzi, F.: "Mixed and nonconforming finite element methods: implementation, postprocessing and error estimates", RAIRO 19 (1985) pp. 7-32.

[2] Arnold, D. N., Douglas, J., Gupta, C. P.: "A family of higher order mixed finite element methods for plane elasticity", Numer. Math. 45 (1984) pp. 1-22.

[3] Babuška, T., Osborn, J., Pitkäranta, J.: "Analysis of mixed methods using mesh dependent norms", Math. Comp. 35 (1980) pp. 1039-1062.

[4] Brezzi, F., Douglas, J., Marini, L. D.: "Two families of mixed finite element methods for second order elliptic problems", Numer. Math. 47 (1985) pp. 217-235.

[5] Ciarlet, P. G.: "The Finite Element Method for Elliptic Problems", North-Holland, Amsterdam 1978.

[6] Johnson, C., Mercier, B.: "Some equilibrium finite element methods for two-dimensional elasticity problems", Numer. Math. 30 (1978) pp. 103-116.

[7] Karp, S.N., Karal, F.C.: "The elastic field in the neighbourhood of a crack of arbitrary angle", CPAM 15 (1962) pp. 413-421.

[8] Pitkäranta, J., Stenberg, R.: "Analysis of some mixed finite element methods for plane elasticity equations", Math. Comp. 41 (1983) pp. 399-423.

[9] Stenberg, R.: "On the construction of optimal mixed finite element methods for the linear elasticity problem", Numer. Math. (to appear).

[10] Williams, M. L.: "Stress singularities resulting from various boundary conditions in angular corners of plates in extension", J. Appl. Mech. 19 (1952) pp. 526-528.

AN EXTENSION THEOREM FOR FINITE ELEMENT SPACES WITH THREE APPLICATIONS

Olof B. Widlund
Courant Institute of Mathematical Sciences
251 Mercer Street
New York, NY 10012 USA

SUMMARY

In this paper, an extension theorem, similar to well-known results for Sobolev spaces, is established for general conforming finite element spaces. Results of this type provide central tools for the theory of iterative substructuring (domain decomposition) and capacitance matrix methods. The main theorem has previously been established, using other techniques, for certain Lagrangian finite elements in the plane. Several applications are discussed in which previous results on iterative substructuring and capacitance matrices are extended to three dimensions, non-Lagrangian finite elements and higher order elliptic problems.

1. INTRODUCTION

Recently there has been a considerable interest in the development and study of iterative substructuring methods for elliptic finite element problems; see Bjørstad and Widlund [6, 7], Bramble, Pasciak and Schatz [8, 9], Chan [11], Chan and Reasco [12], Dihn, Glowinski and Périaux [16], Dryja [17,18], Dryja and Proskurowski [19,20], Golub and Mayers [21], Keyes and Gropp [22] and Widlund [29]. Early work in this area is described in Concus, Golub and O'Leary [15], while the paper by Dryja [17] appears to be the first in which an optimal iterative method is described and analyzed. Much of this work has been for finite difference approximations of second order elliptic problems, often on relatively special plane regions. It is the purpose of this paper to go beyond these cases to more difficult and important finite element problems.

When using these algorithms, also known as domain decomposition methods, the discretized elliptic problem is partitioned into subproblems which correspond to non-overlapping subsets of the region. The subproblems are then solved separately, and repeatedly, while the interaction between the subregions is handled by a conjugate gradient or other suitable iterative method. These iterative methods provide interesting alternatives to the standard industrial finite element practice in which not only the stiffness matrices corresponding to the finite element models for the substructures but also the matrices which represent the interaction between the different parts are fully assembled and factored into their Choleski triangular factors; see Bell, Hatlestad, Hausteen and Ar ldsen [2] and Przemieniecki [25]. For a discussion of different algorithms etc. and a general finite element framework, see Bjørstad and Widlund [7].

The fastest among these methods offers considerable advantages even on sequential computers. They also show particular promise for parallel computing since the last stage of a block Gaussian factorization is likely to lend itself less well to parallel architectures than the earlier stages where work can be carried out for the different substructres without any need of communication between them. Work on the actual parallel implementation of these methods is beginning; see Keyes and Gropp [22] for results on an Intel Hypercube. Similarly, variants of methods for general sparse linear systems of algebraic equations, such as nested dissection algorithms, could be developed in which the factorization would be stopped at a suitable stage. The remaining reduced system, corresponding to the variables not eliminated, would then be solved by an iterative method.

The rate of convergence of iterative methods of this kind can of course be studied by conducting systematic numerical experiments. For general sparse or band matrices it is unlikely that strong general convergence results for powerful preconditioners will be developed for general systems of equations. When a problem arises from the discretization of an elliptic problem additional tools from mathematical analysis are available and the possibility of a systematic development of good preconditioners can become a reality. Thus for an elliptic problem with constant coefficients on a uniform mesh and a region which is a union of a few rectangles, Fourier series can be used; see e.g. section 4 of Bjørstad and Widlund [7]. Similar techniques are used in Bjørstad [4,5] to study a biharmonic problem and in Chan [10] to develop and analyze new fast methods for queuing networks. However, a systematic theory for elliptic problems on general regions requires the development of finite element analogues of certain regularity results for inhomogeneous elliptic boundary value problems, see Bjørstad and Widlund [6, 7] and Bramble, Pasciak and Schatz [8, 9]. The technical aspects of this work is complicated by the fact that the boundaries of the subregions necessarily have corners.

As in the continuous case, we can reduce an elliptic problem with nonhomogeneous boundary data to one which is homogeneous by using an extension theorem; see e.g. Stein [21]. In recent papers, see Bjørstad and Widlund [7] and Bramble, Pasciak and Schatz [8], an extension theorem has been developed for conforming Lagrangian finite elements in the plane. This proof, which essentially has been known at least since 1980, uses a variant of the Bramble-Hilbert lemma; see Ciarlet [13], and a regularity theorem for elliptic differential equations. It appears difficult to extend this proof to non-Lagrangian finite elements, three dimensions and higher order equations since it relies on the boundedness, in a

Sobolev space, of the finite element interpolation operator. The choice of this Sobolev space is limited by the lack of regularity of the solutions of elliptic problems on regions with corners. In this paper, we therefore proceed differently, inspired by an idea used by Astrakhantsev [1] to establish that a capacitance matrix method for a rather special variational difference scheme is optimal for the Neumann boundary condition.

We note that the extension theorems are related to questions concerning the approximation order of finite element spaces for functions which are not sufficiently smooth. These issues were discussed early by Strang [28] for cases without boundaries. This work was extended to some extent by Clément [14] and later quite systematically by Bernardi [3]. While Bernardi carefully includes boundary conditions in her work, her work does not include Hermitian finite elements and it is therefore not directly applicable to all the problems at hand.

In section 3, we discuss three applications. We thus extend Astrakhantsev's result on capacitance matrix methods for Neumann problems to a much more general family of finite elements. While, at least in the West, iterative substructuring methods are attracting much more attention than capacitance matrix algorithms the two families of methods are nevertheless closely related. We also give two new results on domain decomposition in this last section.

2. THE EXTENSION THEOREM

Let Ω be an open, bounded region in R^d, $d \geq 2$, with a piecewise smooth, uniformly Lipschitz continuous boundary Γ. It is well known, cf. e.g. Stein [27], Chapter VI, that there exists a linear operator $\mathcal{E}$ extending functions on Ω to functions on R^d with the properties

i) $\mathcal{E}u|_{\Omega} = u$,

i.e. $\mathcal{E}$ is an extension operator,

ii) $$||\mathcal{E}u||_{H^m(R^d)} \leq C(\Omega)\,||u||_{H^m(\Omega)}, \qquad (2.1)$$

i.e. $\mathcal{E}$ maps $H^m(\Omega)$ continuously into $H^m(R^d)$. m is a nonnegative integer.

The constant $C(\Omega)$ depends on d, m and the Lipschitz constant of the region only. The Sobolev space $H^m(\Omega)$ is the subspace of $L_2(\Omega)$ such that

$$||u||_{H^m(\Omega)} = \left\{ \int_{\Omega} \sum_{|\alpha|\leq m} |(\tfrac{\partial}{\partial x})^{\alpha} u|^2 \, dx \right\}^{1/2} < \infty ,$$

with $(\frac{\partial}{\partial x})^{\alpha} = (\frac{\partial}{\partial x_1})^{\alpha_1} \ldots (\frac{\partial}{\partial x_d})^{\alpha_d}$ and $|\alpha| = \sum \alpha_i$.

We will develop an analogue of this extension theorem for finite element spaces. Let τ^h be a triangulation of Ω into elements. For convenience we assume that each element K is a simplex with sufficiently smooth faces; our theory could equally well be developed for quadrilaterals. The simplices can be curved. As usual we assume that the simplices are properly joined which means that the intersection of any two of them is empty or an entire face of dimension d-1 or less. By h_K we denote the diameter of K and by ρ_K the diameter of the largest ball contained in the element. Following Bernardi [3], we assume that each element is the image of a straight d-simplex $\hat{K}$ under a C^1 map F_K of the form

$$F_K = \tilde{F}_K + \Phi_K$$

where $\tilde{F}_K$ is an invertible affine mapping

$$\tilde{F}_K: \hat{x} \to \tilde{B}_K\hat{x} + \tilde{b}_K$$

and Φ_K satisfies

$$c_K = \sup_{\hat{x}\in\hat{K}} ||D\Phi_K(\hat{x})\hat{B}_K^{-1}|| \le c < 1$$

with c uniformly bounded away from 1. We note that $\tilde{K} = \tilde{F}_K(\hat{K})$ is a straight d-simplex and that we can arrange it so that K and $\tilde{K}$ have the same vertices, i.e. $\tilde{K}$ is a straight approximation of K. Isoparametric elements as well as those based on an exact triangulation, cf. Bernardi [3], Lenoir [23] and Scott [26], can easily be accommodated in this framework. The triangulation is assumed to be regular, i.e. $\tilde{h}_K/\tilde{\rho}_K \le \sigma$, where σ is uniformly bounded for all h. These assumptions exclude degenerate, very flat elements and also lead to a bound on the number of simplices that have a particular vertex in common, but still allow us selectively to refine the triangulation to improve the accuracy where the solution is less smooth. In practice a simplex will be straight, unless at least two of its vertices fall on the boundary or an interface between subregions modeled separately e.g. because of different material properties. We will not discuss the effects of approximating Ω by a union of straight of isoparametric elements but always assume that the triangularions of Ω are exact.

We will consider general, conforming finite elements, i.e. the approximating space $V^h \subset V$ where V is the linear space appropriate for the elliptic variational problem to be considered; see Ciarlet [13]. A finite element is defined on the element level by a triple (K,P_K,Σ_K) where K is the simplex and P_K a space of functions. The dimension of P_K is bounded uniformly. Normally P_K is the image under the map F_K of a space of polynomials. We note that these functions can be vectored values. No essential difficulties are introduced by such an assumption nor by considering curved rather than straight simplices. Σ_K is a set of linear functionals defined on smooth functions and on P_K such that the related interpolation problem on K is uniquely solvable in P_K.

From the set $\bigcup_{K\in\tau_h} \Sigma_K$ a maximal system of linearly independent linear functionals on $C^\infty(\bar{\Omega})$ is extracted. Basis functions for the finite element space V^h can then be constructed as e.g. in Ciarlet [13], creating $\{\phi_i\}_{i=1}^{N_h}$ and $\{\mu_i\}_{i=1}^{N_h}$ which are a dual pair of bases for the space V^h and the space of linear functionals, respectively i.e.

$$\mu_i(\phi_j) = \delta_{ij}\ , \quad 1 \le i \le N_h\ , \quad 1 \le j \le N_k\ .$$

For $1 \le i \le N_h$, we define

$$\Delta_i = \bigcup_{K\in\tau^h} \{K;\ \text{supp}\ \mu_i \text{ overlaps } K\}\ .$$

Under the assumptions given above one can show straightforwardly that the number of elements that form Δ_i is uniformly bounded and that the diameters of any elements K and K', contained in the same Δ_i satisfy

$$\tilde{h}_K \le C\tilde{h}_{K'}\ ,$$

where C depends on $c = \max c_K$ and $\sigma = \max \tilde{h}_K/\tilde{\rho}_K$ only. The support of ϕ_i is contained in Δ_i .

In our proof we assume, following Strang [28], that the basis functions $\phi_i(x)$ are uniform of order m, i.e. there exists constants c_s , independent of h and i, such that

$$\max_{\substack{x\in\Delta_i\\ |\alpha|=s.}} \left|\left(\frac{\partial}{\partial x}\right)^\alpha \phi_i(x)\right| \le c_s\, h^{d_i-s}\ , \quad s \le m\ . \tag{2.2}$$

We note that d_i often is the degree of a derivative associated with the degree of freedom in question.

Assuming that $C\Omega$, the complement of Ω,has been triangulated in an equally benign way and that $V^h(\Omega)$ has been extended to $V^h(R^d) \subset H^m(R^d)$, we are ready to describe how to find an extension $\mathcal{E}^h u_h \in V^h(R^d)$ of a given element $u_h \in V^h(\Omega)$ such that

$$||\mathcal{E}^h u_h||_{H^m(R^d)} \le C||u_h||_{H^m(\Omega)}\ . \tag{2.3}$$

In a first step, we use the original extension theorem and the fact that the space is conforming, i.e. $u_h \in H^m(\Omega)$, to find $\mathcal{E}u_h \in H^m(R^d)$ such that

$$||\mathcal{E}u_h||_{H^m(R^d)} \le C(\Omega)||u_h||_{H^m(\Omega)}\ . \tag{2.4}$$

However $\mathcal{E}u_h \notin V^h(R^d)$.

In a second step, we follow Strang [25] and smooth $\mathcal{E}u^h$ and interpolate to find an element $w_h \in V^h(R^d)$ such that

$$||w_h||_{H^m(R^d)} \le C||\mathcal{E}u_h||_{H^m(R^d)} \le C(\Omega)C\,||u_h||_{H^m(\Omega)}\ . \tag{2.5}$$

However $w_h|_\Omega \ne u_h$ and w_h is therefore not an extension of u_h. We note that Strang [28] shows that

$$||w_h - \mathcal{E}u_h||^2_{L^2(K)} \leq Ch_K^{2m} \sum_{K'}^{fin} ||w_h - \mathcal{E}u_h||^2_{H^m(K')} \quad . \qquad (2.6)$$

where the sum is over a fixed number of neighboring elements.

In a third step, we patch w_h and u_h together to create $\mathcal{E}^h u_h$. We interpolate using all the original parameters obtained from u_h in $\overline{\Omega}$ and in $C\overline{\Omega}$, the complement of $\overline{\Omega}$, those associated with w_h. The resulting function $\mathcal{E}^h u_h \in V^h(R^d)$. For any element $K \subset \Omega$ the original values of u_h are recovered. We also note that similarly, in $C\Omega$, $\mathcal{E}^h u_h$ differs from w_h only on elements which have at least one vertex on Γ. There remains to establish (2.3).

By the triangle inequality

$$||\mathcal{E}^h u_h||_{H^m(C\Omega)} \leq ||\mathcal{E}^h u_h - w_h||_{H^m(C\Omega)} + ||w_h||_{H^m(C\Omega)} \quad .$$

By (2.1) and (2.5), $||w_h||_{H^m(C\Omega)}$ can be estimated by $||u_h||_{H^m(\Omega)}$.

Consider an element $K' \subset C\Omega$, with at least one vertex on Γ for which $\mathcal{E}^h u_h - w_h$ is not identically zero. The parameters associated with the local interpolation problem are equal to zero except for those shared with at least one element $K \subset \Omega$. By using (2.2) one can establish that

$$||\mathcal{E}^h u_h - w_h||^2_{H^m(K')} \leq ch_K^{-2m} ||\mathcal{E}^h u_h - w_h||^2_{L^2(K')} \quad . \qquad (2.7)$$

Under the regularity assumptions introduced above and (2.2), we can estimate this L_2 norm by $h_K^{d/2}\times$the ℓ_2 norm of the vector of parameters $h^{d_i}\mu_i(\mathcal{E}^h u_h - w_h)$. When considering the corresponding norms of $\mathcal{E}^h u_h - w_h$ on the neighboring element $K \subset \Omega$, we see that

$$||\mathcal{E}^h u_h - w_h||_{L^2(K')} \leq C||\mathcal{E}^h u_h - w_h||_{L^2(K)} = C||\mathcal{E}u_h - w_h||_{L^2(K)} \quad .$$

The crucial observation is that this inequality holds whatever the values of the additional parameters associated with K happen to be.

The proof is concluded by adding over all the relevant elements of $C\Omega$ and using inequalities (2.6) and (2.7). An exact analogue of the extension theorem given in the beginning of this section has thus been established for conforming finite element spaces defined on sufficiently regular triangulations. We note that $\mathcal{E}^h$ is a linear operator and if desired we can modify $\mathcal{E}^h u_h$ so that it vanishes at a fixed distance outside Γ.

3. APPLICATIONS

In this section, we will briefly discuss the application of the extension theorem to capacitance matrix and iterative substructuring methods. We note that the main idea behind the proof of our first result is due to Astrakhantsev [1].

There are occasions when a linear system of equations can be imbedded in a larger system which is easier to solve. For the expanded system an efficient preconditioner might be available e.g. if the geometry of the corresponding region is rectangular and fast Poisson solvers can be used; see Proskurowski and Widlund [24] and the references therein.

Here we will only consider the solution of Neumann problems for selfadjoint elliptic problems by capacitance matrix methods. The region Ω is imbedded in Λ, a rectangle or some other convenient, simple region. The boundaries of Ω and Λ do not intersect and the two regions are triangulated as in section 2. A symmetric bilinear form $a_\Omega(u,v)$ is associated with the original elliptic problem. To avoid nonessential details, we assume that the problem is strictly elliptic in the sense that

$$c||u||^2_{H^m(\Omega)} \leq a_\Omega(u,u) .$$

We also assume that this bilinear form can be extended to a similarly well behaved form $a_\Lambda(u,v)$. We note that we have considerable freedom in choosing a boundary condition on the boundary of Λ.

The space $H^m(\Lambda)$, or an appropriate subspace of functions satisfying these boundary conditions, is approximated by $\tilde{V}^h(\Lambda)$, a conforming finite element space satisfying the conditions of section 2. The restriction $V^h(\Omega)$ of this space to Ω is used to solve the Neumann problem on Ω. We note that all the parameters associated with $V^h(\Omega)$ are determined when the discrete Neumann problem is solved. Denote by A the resulting positive definite, symmetric stiffness matrix and by n_h the order. The matrix corresponding to the problem on the larger region Λ is denoted by B and its order is N_h.

One can think of the capacitance matrix method as a preconditioned conjugate gradient method in which the trivially expanded system

$$\begin{pmatrix} A & 0 \\ 0 & 0 \end{pmatrix} \begin{pmatrix} x_1 \\ x_2 \end{pmatrix} = \begin{pmatrix} b_1 \\ 0 \end{pmatrix}$$

is solved using B as a preconditioner. As always the convergence of the method is determined by the spectrum of

$$\begin{pmatrix} A & 0 \\ 0 & 0 \end{pmatrix} \phi = \lambda B \phi$$

or equivalently by the stationary values of the Rayleigh quotient

$$\frac{x_1^T A x_1}{x^T B x} .$$

Let $\lambda_1 \geq \lambda_2 \geq \ldots$ be the eigenvalues. We note that

$\lambda_{n_h+1} = \lambda_{n_h+2} = \ldots = \lambda_{N_h} = 0$ and that $\lambda_1 \le 1$ since the strain energy on Λ is larger than that on the smaller region Ω. The null space of dimension $N_h - n_h$ does not affect the algorithm, cf. Astrakhantsev [1] or Proskurowski and Widlund [24]. To prove that the rate of convergence is independent of the mesh size we must establish a good lower bound on λ_{n_h}. Such a bound is obtained by the Courant-Fischer theorem. The $N_h - n_h$ linear constraints used are obtained by using the extension theorem. The value of each degree of freedom associated with $\Lambda \setminus \Omega$ is, by the linearity of $\mathcal{E}^h$, a linear combination of those associated with $V^h(\Omega)$.

The extension theorem therefore provides a bound of the strain energy on Λ in terms of that on Ω and the proof is completed. We note that numerical experiments reported in Proskurowski and Widlund [24] show that this bound, and thus the constant in the extension theorem, deteriorates if triangles become very thin.

We next turn to a discussion of an optimal iterative substructuring method; cf. Bjørstad and Widlund [7] for a detailed discussion of this and similar methods. We begin by discussing Lagrangian finite elements, extending our previous results to dimensions higher than two. Since Lagrangian elements do not belong to H^2, we confine ourselves to second order systems. To simplify the notations, we also concentrate on the case where the original region Ω is the union of two regions Ω_1 and Ω_2 and Γ_3, the intersection of the closures of Ω_1 and Ω_2. The boundaries of Ω_1, Ω_2 and Ω are $\bar{\Gamma}_1 \cup \bar{\Gamma}_3$, $\bar{\Gamma}_2 \cup \bar{\Gamma}_3$ and $\bar{\Gamma}_1 \cup \bar{\Gamma}_2$ respectively. We will work with a Dirichlet condition on the boundary of Ω.

The variational formulation is then

$$a_\Omega(u_h, v_h) = f(v_h), \quad \forall\, v_h \in V_0^h(\Omega) \; . \tag{3.1}$$

$$u_h \in V^h(\Omega), \; u_h \text{ given on } \bar{\Gamma}_1 \cup \bar{\Gamma}_2 \, ,$$

where $V_0^h(\Omega)$ is the subspace of functions in $V^h(\Omega)$ which vanish on $\bar{\Gamma}_1 \cup \bar{\Gamma}_2$. We again assume that $a_\Omega(u,v)$ is uniformly elliptic and selfadjoint.

In the proof, we need to extend an element of $V_0^h(\Omega_1, \Gamma_1)$, the subspace of $V^h(\Omega_1)$ of functions vanishing on Γ_1, to an element of $V_0^h(\Omega)$. We can accomplish this by extending the given element by zero in $C\Omega$ and then use the construction of section 2 to obtain values in Ω_2.

The linear system of equations corresponding to (3.1) has the form

$$Kx = \begin{pmatrix} K_{11} & 0 & K_{13} \\ 0 & K_{22} & K_{23} \\ K_{13}^T & K_{23}^T & K_{33} \end{pmatrix} \begin{pmatrix} x_1 \\ x_2 \\ x_3 \end{pmatrix} = \begin{pmatrix} b_1 \\ b_2 \\ b_3 \end{pmatrix} \tag{3.2}$$

where K is positive definite and symmetric. The matrix K_{11} represents the couplings between pairs of degrees of freedom in Ω_1, K_{13} couplings between pairs belonging to Ω_1 and Γ_3 respectively etc.

We make the assumptions that the discrete problems on the subregions, with some appropriate boundary condition added on Γ_3, can be solved exactly. We can therefore set b_1 and b_2 equal to zero. By block Gaussian elimination, the problem (3.2) is reduced to

$$Sx_3 = (K_{33} - K_{13}^T K_{11}^{-1} K_{13} - K_{23}^T K_{22}^{-1} K_{23}) x_3 = b_3 \,. \tag{3.3}$$

This system is preconditioned by $S^{(1)}$, a matrix related to a problem on Ω_1 as follows. By using a natural boundary condition on Γ_3 as the only nonhomogeneous data and a zero Dirichlet condition on Γ_1, we obtain

$$\begin{pmatrix} K_{11} & K_{13} \\ K_{13}^T & K_{33}^{(1)} \end{pmatrix} \begin{pmatrix} x_1 \\ x_3 \end{pmatrix} = \begin{pmatrix} 0 \\ c_3 \end{pmatrix} , \tag{3.4}$$

where the elements of $K_{33}^{(1)}$ have the form $a_{\Omega_1}(\phi_i, \phi_j)$, with ϕ_i, ϕ_j basis functions associated with degrees of freedom on Γ_3.

The problem reduces to

$$S^{(1)} x_3 = (K_{33}^{(1)} - K_{13}^T K_{11}^{-1} K_{13}) x_3 = c_3 \,.$$

Since $S = S^{(1)} + S^{(2)}$, where $S^{(2)}$ is constructed in the same way as $S^{(1)}$, it is not surprising that $S^{(1)}$ is an excellent preconditioner of S.

To provide upper and lower bounds for the associated generalized eigenvalue problem, we consider the Rayleigh quotient

$$\frac{x_3^T S^{(1)} x_3}{x_3^T S x_3} \,.$$

We call a solution of (3.4) a discrete harmonic and we similarly say that the restrictions to Ω_1 and Ω_2 of the solution corresponding to (3.2) with zero b_1 and b_2 is a discrete harmonic function on Ω_1 as well as on Ω_2. The numerator of the Rayleigh quotient represents the strain energy of a discrete harmonic function on Ω_1 while the denominator also contains the strain energy of its discrete harmonic extension to Ω_2. An upper bound is therefore trivially obtained. We need to estimate $x_3{}^T S^{(2)} x_3$ in terms of $x_3{}^T S^{(1)} x_3$ to obtain a lower bound. We note that

$$x_3^T S^{(2)} x_3 = a_{\Omega_2}(u_h, u_h)$$

for some $u_h \in V_0^h(\Omega_2, \Gamma_2)$ which also satisfies

$$a_{\Omega_2}(u_h, v_h) = 0 \,, \quad \forall\, v_h \in V_0^h(\Omega_2) \,.$$

Therefore

$$a_{\Omega_2}(u_h+v_h,\ u_h+v_h) = a_{\Omega_2}(u_h,u_h) + a_{\Omega_2}(v_h,v_h)\ ,\quad \forall\, v_h \in V_0^h(\Omega_2),$$

i.e. u_h has the least strain energy of any element in $V_0^h(\Omega_2,\Gamma_2)$ with the given boundary values on Γ_3. In view of this the extension theorem, as modified in this section, can be used to complete the argument.

We conclude with some comments on a non-Lagrangian case. To be specific, we consider the Dirichlet problem for the biharmonic equation approximated by the 18 degrees of freedom, reduced quintic element discussed in Ciarlet [13]. The interpolation conditions are associated with the values of the function and all its first and second derivatives at the vertices of the triangles. Here we principally want to make the point that there are two different ways of obtaining an $H^2(\Omega)$ extension from the subregion Ω_1 to Ω_2 and therefore two domain decomposition algorithms. For simplicity, we assume that Γ_3 is straight.

In the first variant all the degrees of freedom associated with Γ_3 are shared by the finite element functions on Ω_1 and Ω_2. The proof of the optimality of the domain decomposition algorithm carries over directly from the case discussed above. In the iteration, approximate values of six parameters per vertex on Γ_3 are obtained. In the second variant, we keep the values of $\dfrac{\partial^2 u_h}{\partial n^2}$ free when extending the discrete biharmonic function from Ω_1 to Ω_2. Since this corresponds to the removal of constraints we have no further problem establishing the necessary bound. From an algorithmic point of view there is a possible benefit. A detailed examination of the linear algebra involved reveals that we only obtain five nontrivial residuals per vertex on Γ_3, corresponding to the parameters shared between Ω_1 and Ω_2. We note that the values of $\partial^2 u_h/\partial n^2$ at the vertices do not affect the values of u_h and its gradient on Γ_3. A solution obtained by this second variant will thus belong to H^2 but in general it will differ from the standard discrete solution obtained by applying the finite element method on the entire region Ω.

ACKNOWLEDGMENTS

This work was supported by the National Science Foundation under Grant NSF-DCR-8405506, and by the U. S. Department of Energy, under Contract DE-AC02-76ER03077-V, at the Courant Mathematics and Computing Laboratory.

REFERENCES

[1] ASTRAKHANTSEV, G. P.: "Method of fictitious domains for a second-order elliptic equation with natural boundary conditions," U.S.S.R. Computational Math. and Math. Phys. 18 (1978) pp. 114-121.

[2] BELL, K., HATLESTAD, B., HAUSTEEN, O. E. and ARALDSEN, P.O.: "NORSAM, a programming system for the finite element method," User manual, part 1 (1973), Veritas, Høvik, Norway.

[3] BERNARDI, C.: "Optimal finite element interpolation on curved domains," Technical report 84017, Univ. of Paris VI.

[4] BJØRSTAD, P. E.: "Numerical solution of the biharmonic equation," Stanford University thesis, Stanford, CA.

[5] BJØRSTAD, P. E.: "Fast numerical solution of the biharmonic Dirichlet problem on rectangles," SIAM J. Numer. Anal. 20 (1983), pp. 59-71.

[6] BJØRSTAD, P. E. and WIDLUND, O. B.: "Solving elliptic problems on regions partitioned into substructures," Elliptic Problem Solvers, II (Birkhoff, G. and Schoenstadt, A., eds.), Academic Press, New York (1984), pp. 245-256.

[7] BJØRSTAD, P. E. and WIDLUND, O. B.: "Iterative methods for the solution of elliptic problems on regions partitioned into substructures," SIAM J. N.A. (1986), to appear.

[8] BRAMBLE, J. M., PASCIAK, J. E. and SCHATZ, A. M.: "An iterative method for elliptic problems on regions partitioned into substructures," Math. Comp. 46 (1986) pp. 361-370.

[9] BRAMBLE, J. H., PASCIAK, J. E. and SCHATZ, A. M.: "The construction of preconditioners for elliptic problems by substructuring," Math. Comp., to appear.

[10] CHAN, R. H.: "Iterative methods for overflow queuing models," New York University, thesis, August 1985.

[11] CHAN, T. F.: "Analysis of preconditioners for domain decomposition," Yale Univ. Computer Science Dept. Technical Report RR-408 (1985).

[12] CHAN, T. F. and REASCO, D. C.: "A survey of preconditioners for domain decomposition," Yale Univ. Computer Science Dept. Tech. Report RR-414 (1985).

[13] CIARLET, P. G.: "The finite element method for elliptic problems," North-Holland (1978).

[14] CLÉMENT, Ph.: "Approximation by finite element functions using local regularization," R.A.I.R.O. R-9 (1975) pp. 77-84.

[15] CONCUS, P., GOLUB, G. H. and O'LEARY, D. P.: "A generalized conjugate gradient method for the numerical solution of elliptic PDEs," Sparse matrix computations (Bunch, J.R. and Rose, D.F., eds.), Academic Press, New York, 1976, pp. 309-332.

[16] DIHN, Q. V., GLOWINSKI, R. and PÉRIAUX, J., "Solving elliptic problems by domain decomposition with applications," Elliptic problem solvers II (Birkhoff, G. and Schoenstadt, A., eds.), Academic Press, New York, 1984, pp. 395-426.

[17] DRYJA, M.: "A capacitance matrix method for Dirichlet problem on polygon region," Numer. Math. 39 (1982), pp. 51-64.

[18] DRYJA, M.: "A finite element-capacitance method for elliptic problems on regions partitioned into substructures," Numer. Math. 44 (1984), pp. 153-168.

[19] DRYJA, M. and PROSKUROWSKI, W.: "Fast elliptic solvers on rectangular regions subdivided into strips," Adv. Computer Meth. for PDE, V, Vichnevetsky, R. and Stapleman, R. S., eds.) IMACS (1984), pp. 360-368.

[20] DRYJA, M. and PROSKUROWSKI, W.: "Capacitance matrix method using strips with alternating Neumann and Dirichlet boundary conditions," Appl. Numer. Math. 1 (1985), pp. 285-298.

[21] GOLUB, G. H. and MAYERS, D.: "The use of preconditioning over irregular regions," Proceedings from the Sixth International Conference on Computing Methods in Science and Engineering, Versailles, France, Dec. 12-16, 1983.

[22] KEYES, D. E. and GROPP, W. D.: "A comparison of domain decomposition techniques for elliptic partial differential equations and their parallel implementation," to appear in SIAM J. Sci. Stat. Comput.

[23] LENOIR, M.: "Optimal isoparametric finite elements and error estimates for domains involving curved boundaries," to appear.

[24] PROSKUROWSKI, W. and WIDLUND, O.: "A finite element - capacitance matrix method for the Neumann problem for Laplace's equation," SIAM J. Sci. Stat. Comput. 1 (1980), pp. 410-425.

[25] PRZEMIENIECKI, J. S.: "Matrix structural analysis of substructures," Am. Inst. Aero. Astro. J. 1 (1963), pp. 138-147.

[26] SCOTT, R.: "Finite element techniques for curved boundaries," Thesis, MIT (1973).

[27] STEIN, E. M.: "Singular integrals and differentiability properties of functions," Princeton Univ. Press (1970).

[28] STRANG, G.: "Approximation in the finite element method," Numer. Math. 19 (1972), pp. 81-98.

[29] WIDLUND, O. B.: "Iterative methods for elliptic problems on regions partitioned into substructures and the biharmonic Dirichlet problem," Proceedings from the Sixth International Conference on Computing Methods in Science and Engineering," Versailles, France, Dec. 12-16 (1983).

List of lectures presented at the seminar

O. AXELSSON (Nijmegen): Some new applications of a mixed variable finite element method for the efficient solution of nonlinear diffusion problems

F. BREZZI (Pavia): Finite element method for Mindlin-Reissner plates

Y. EGUCHI, L. FUCHS (Stockholm): A conjugate gradient finite element scheme for Navier-Stokes equations

K. ERIKSSON (Göteborg): Adaptive finite element method for elliptic and parabolic problems [1]

L. FUCHS (Stockholm): An efficient numerical scheme for transonic vortical flows [1]

F.K. HEBEKER (Paderborn): On the numerical treatment of viscous flows against bodies with corners and edges by boundary element and multigrid methods [1]

H. JARAUSCH, W. MACKENS (Aachen): Computing nondegenerate bifurcation points by an adaptive numerical Lyapunov-Schmidt reduction

C. JOHNSON (Göteborg): Convergence of a finite element method for a nonlinear conservation law

B. KRÖPLIN (Dortmund): A technique for structural instability analysis [1]

P. LETALLEC (Paris): Numerical decomposition methods for adhesion problems in finite elasticity [1]

W. MACKENS, H. JARAUSCH (Aachen): Computing bifurcation diagrams for large nonlinear variational problems

H. MITTELMANN (Tempe, USA): Multi-grid continuation [1]

P. NEITTAANMÄKI (Jyvaskyla): Post-processing for a finite element-scheme with linear elements [1]

Z.P. NOWAK (Kiel): Higher order panel method for potential flow problems in 3D [2]

J. PITKÄRANTA (Espoo): On simple finite element method for Mindlin plates [1]

R. RANNACHER (Saarbrücken): On Richardson's extrapolation for finite element method [1]

H. SCHWETLIK (Halle): Higher order predictors and adaptive stepsize control in path following algorithms

ZHON-CI SHI (Hefei): On the convergence behaviour of Zienkiewicz cubic plate element and its modification

R. STENBERG (Helsinki): On the mixed finite element method for the elasticity problem [1]

W. WENDLAND (Darmstadt): Zur asymptotischen Konvergenz kombinierter FEM-BEM-Methoden

O.B. WIDLUND (New York): An extension theorem for finite element spaces with three applications [1]

Further participants:

H.W. Alt (Bonn), D. Bischoff (Hannover), H. Blum (Saarbrükken), D. Braess (Bochum), G. Brand (Hannover), J. Dehnhardt (Hannover), E. Martensen (Karlsruhe), G. Dzink (Bonn), J. Feldermann (Aachen), G. Hofmann (Kiel), H. Lent (Hannover), J.F. Maitre (Lyon), Obrecht, R. Pallacks (Hamburg), P. Peisker (Bochum), A. Reed (Oxford), J. Roux (Clamart, F), Seeber (Loeben), R. Verfürth (Heidelberg), C. Wächter (Aachen), B. Werner (Hamburg), G. Wittum (Kiel), W. Wunderlich (Bochum), H. Yserentant (Dortmund)

---

1) written version contained in these proceedings

2) published in "Finite Approximation in Fluid Mechanics" (E.H.Hirschel, ed.), Volume 14 of Notes on Numerical Fluid Mechanics, Vieweg, Braunschweig/Wiesbaden, 1986, pp. 218-231.